Powering Through: Energy Resilience Planning from Grid to Government

RIVER PUBLISHERS SERIES IN ENERGY MANAGEMENT

The "River Publishers Series in Energy Management" is a series of comprehensive academic and professional books focussing on management theory and applications for energy related industries and facilities. Books published in the series serve to provide discussion and exchange information on management strategies, techniques, methodologies and applications, with a focus on the energy industry.

Topics include management systems, handbooks for facility management, safety, security, industrial strategies, maintenance and financing, impacting organizational communications, processes and work practices. Content is also featured for energy resilient and high-performance buildings.

The main aim of this series is to serve as a useful reference for academics, researchers, managers, engineers, and other professionals in related matters with energy management practices.

Books published in the series include research monographs, edited volumes, handbooks and textbooks. The books provide professionals, researchers, educators, and advanced students in the field with an invaluable insight into the latest research and developments.

Topics covered in the series include, but are not limited to:

- Facility management;
- Safety and security;
- Management systems and solutions;
- Industrial energy strategies;
- Financing and costs;
- Energy resilient buildings;
- Green buildings management.

For a list of other books in this series, visit www.riverpublishers.com

Powering Through: Energy Resilience Planning from Grid to Government

Alex Rakow & Brian Levite

River Publishers

Routledge
Taylor & Francis Group
NEW YORK AND LONDON

Published 2023 by River Publishers
River Publishers
Alsbjergvej 10, 9260 Gistrup, Denmark
www.riverpublishers.com

Distributed exclusively by Routledge
605 Third Avenue, New York, NY 10017, USA
4 Park Square, Milton Park, Abingdon, Oxon OX14 4RN

Powering Through: Energy Resilience Planning from Grid to Government /
by Alex Rakow & Brian Levite.

Routledge is an imprint of the Taylor & Francis Group, an informa business

ISBN 978-87-7022-786-5 (print)
ISBN 978-10-0084-852-6 (online)
ISBN 978-10-0337-491-6 (ebook master)

Contents

List of Figures **ix**

List of Tables **xi**

List of Abbreviations **xiii**

Introduction **1**

Part 1: Energy Resilience Principles ***7***

1 The Energy Resilience Imperative **9**

Access to Fossil Fuels *9*

The Energy Water Nexus. *13*

Electricity Infrastructure. *16*

Terrorism and Sabotage *20*

Climate Change *23*

2 Energy Resilience Concepts **31**

The Evolving Meaning of Energy Resilience. *31*

Resilience at all Levels *33*

Local Energy Resilience *33*

Grid Resilience *35*

Government Resilience. *36*

Finding a place on the Energy Resilience Spectrum *38*

Part 2: Energy Resilience Strategies and Tactics ***43***

3 The Energy Resilience Planning Process **45**

Rallying Around Public Safety. *47*

Stressing the Green Benefits: Environmental and Financial *48*

Using a Mandate to Leverage Better Resilience Planning *48*

Step 1: Engage Stakeholders and Establish Scope 49
Step 2: Assess Energy Resilience Baseline 51
Step 3: Perform Risk Assessment 52
Step 4: Set Goals and Identify Resilience Tactics 53
Step 5: Plan Implementation of Resilience Tactics 57
Step 6: Execute Measures and Evaluate Performance 62
Resilience Planning for the Grid 63
Resilience Planning for Government 65
Resilience Planning for Local Institutions 68

4 Resilience Metrics and Maturity Model **71**
The Maturity Model Approach 73
The Resilience Value Approach 76
The Customized Metric Approach 78
Getting Started with the Maturity Model 80

Part 3: Energy Resilience at Every Level **83**

5 Energy Resilience at the Grid Level **85**
The Changing Needs of the Grid 85
The Increasing Cost of Outages 85
The Potential of Distributed Energy Resources 86
The Demands of Electrification 87
The Resilient Grid is a Smart Grid 88
Best Practices for Utility Resilience Planning 89
What Can Utilities do to Improve Resilience? 93
Utility Resilience Program Maturity Model 107

6 Energy Resilience at the Local Level **109**
The Importance of Local Energy Resilience 109
Energy Resilience in our Four Community Types 110
Local Resilience Special Cases: Cities 123
Local Resilience Special Cases: Portfolios of Distant Facilities 127

Tactical Approaches to Local Energy Resilience *129*
On-Site Energy Technologies *138*
Local Energy Resilience Maturity Model *153*

7 Government Resilience: Policy and Programs **157**
Government Actions in Support of Government Energy Resilience . . . *158*
Government Actions to Support Market Energy Resilience *161*
Government Resilience Program Maturity Model *172*

Conclusion **175**

Index **179**

About the Authors **183**

Endnotes **185**

List of Figures

Figure 1.1	Data from "Estimated use of water in the United States in 2015." Dieter et. al. USGS. Circular 1441. 2018..	14
Figure 1.2	Climate Central. *Power OFF: Extreme Weather and Power Outages.* Sept. 30, 2020. ClimateCentral.org.	27
Figure 1.3	"Electricity Substation Facilities on the Gulf Coast Exposed to Storm Surge from Category 1 Hurricanes under Four Increments of Future Sea Level Rise." Oak Ridge National Laboratory, 2015.	28
Figure 2.1	A. Bloom et al., "The Value of Increased HVDC Capacity Between Eastern and Western U.S. Grids: The Interconnections Seam Study." IEEE Transactions on Power Systems. vol. 37, no. 3. May 2022..	38
Figure 2.2	The spectrum of energy resilience for local communities or institutions.	39
Figure 3.1	Energy resilience plan structure.	55
Figure 3.2	Green revolving fund.	61
Figure 3.3	Steps for conducting a vulnerability assessment and developing climate resilience solutions.	65
Figure 3.4	Community Energy Strategic Planning Timeline.	66
Figure 3.5	Resilience Framework Process, US Dept. of Homeland Security.	66
Figure 3.6	Technical Resilience Navigator Modules.	69
Figure 4.1	Qualitative resilience evaluation example.	72
Figure 4.2	Quantitative performance assessment example.	72
Figure 4.3	DHS site resilience checklist example from their Resilience Framework plan..	75
Figure 4.4	Schematic of the GMLC's integrated resilience approach.	78
Figure 4.5	GMLC's inputs and outputs table for the two analytic spirals.	78

Figure 5.1 "Electric Grid Resilience Solutions." S&C Electric Company, 2021. 94

Figure 5.2 "Radial vs. Networked Grid Topology." S&C Electric Company, 2021. 100

Figure 6.1 Office of the Assistant Secretary of Defense for Sustainment, Energy Resilience Program. "Energy Resilience Timeline." U.S. Department of Defense, 2019.. 115

Figure 6.2 Example of a microgrid system. 152

Figure 7.1 Characteristics of solution commercialization. 162

Figure 7.2 A lab in the NREL ESIF facility simulating energy usage across multiple buildings. "Energy-Efficient and Grid-Interactive Buildings." National Renewable Energy Laboratory. 2022... 165

List of Tables

Table 3.1 Commonalities in recent federal government resilience guides 46

Table 3.2 Continuity criticality quantitative scoring definitions (Table D-2 from *DHS Resilience Framework*) 52

List of Abbreviations

ASCE	American Society of Civil Engineers
ASHRAE	American Society of Heating, Refrigeration and Air Conditioning Engineers
CPUC	California Public Utility Commission
CHP	Combined heat and power
CBA	Cost–benefit analysis
COS	Cost-of-service
DOE	Department of Energy
DERs	Distributed energy resources
EV	Electric vehicle
EaaS	Energy-as-a-Service
ECM	Energy conservation measure
ERCIP	Energy Resilience and Conservation Investment Program
EPA	Environmental Protection Agency
FERC	Federal Energy Regulatory Commission
FP&L	Florida Power & Light
GHG	Greenhouse gas
GMLC	Grid Modernization Laboratory Consortium
HMGP	Hazard Mitigation Grant Program
IPPs	Independent power producers
IIJA	Infrastructure Investment and Jobs Act
MCDA	Multi-criteria decision analysis
NARUC	National Association of Regulatory Utility Commissioners
NASEO	National Association of State Energy Officials
NREL	National Renewable Energy Laboratory
NRC	National Research Council

NYSERDA	New York State Energy Research and Development Authority
NYRev	New York State's Reforming the Energy Vision
NERC	North American Electric Reliability Council
OPEC	Organization of Petroleum Exporting Countries
PBR	Performance-based regulation
PEER	Performance Excellence in Electricity Renewal
PTC	Production tax credit
PSPS	Public Safety Power Shutoffs
RISE	Resilience Innovation for a Stronger Economy
SGIG	Smart Grid Investment Grant
SPIDERS	Smart Power Infrastructure Demonstration for Energy Reliability and Security
SCADA	Supervisory control and data acquisition
SAIDI	System Average Interruption Duration Index
SAIFI	System Average Interruption Frequency Index
TEPCO	Tokyo Electric Power Company Holdings Inc.
TCDR	Town Center Distributed Energy Resources
UPS	Uninterruptable power supply
DHS	U.S. Department of Homeland Security

Introduction

On August 29, 2021 – the 16th anniversary of Hurricane Katrina – Hurricane Ida made landfall in southeastern Louisiana. The storm brought 150 mph winds to the state for only the third time in its history (Katrina topped out at 125 mph over land) and quickly left nearly one million customers without power in Louisiana, Mississippi, and Alabama. The electric utility Entergy reported that the storm damaged more than 30,000 of its distribution poles in the area. As the storm moved northeast, it prompted tornados and flooding from Pennsylvania to Cape Cod, causing outages for another 232,000 customers. Flooding was so sudden and severe in New York that the city was forced to close nearly its entire transportation system. Half a million customers were still without power on September 7 – a week after landfall – and 15% of customers were still waiting a week after that. The storm took many lives, including 10 people in New Orleans who died from heat exposure during the extended power outage, six who died of carbon monoxide poisoning from gas generators, and two utility workers who lost their lives trying to repair downed lines.[1,2]

As severe and widespread as the effects of Ida were, the average American could be forgiven for not recalling much about it unless they were directly impacted. Hurricane Ida was one of 20 separate severe weather events in 2021 to cause over $1 billion in damage in the United States. In the same year, historic cold temperatures reached down from Nebraska into Texas, wreaking havoc for weeks. The western U.S. endured another crushing year of drought, wildfires, and heatwaves. Rare and damaging tornados sprung up repeatedly across the southeast and Midwest. The record tying windspeed that Ida brought to Louisiana had been matched only one year prior by Hurricane Laura in 2020.[3]

The rise of severe weather and natural disasters over the past two decades has wrought a wide variety of damage to infrastructure and public safety. When these events impact our energy system, as they almost invariably do, this damage is multiplied many times over. Those fleeing a flood or wildfire may find that the shelters and hospitals they need are without power. Those whose homes are undamaged by a storm may still find they are exposed to dangerous heat or cold in the days that follow.

Of course, extreme weather is not the only threat facing our energy system. Far-away military conflicts and market manipulation can leave businesses and homeowners at the mercy of wild fluctuations in energy prices. The age and complexity of our energy infrastructure in the United States can turn a single failure into a widespread disruption – a fact that makes our energy system all-the-more attractive to would-be saboteurs, both cyber and analog.

Each of these threats is growing, and every day the task of protecting our energy security grows along with them. In this book, we examine this challenge from the perspective of government regulators, grid operators, and consumers and describe how the pursuit of energy resilience for each group creates a more resilient system for all.

Energy Resilience Over Energy Independence

The concept of energy independence is a durable one – a perennial campaign promise meant to evoke a future in which our energy security is assured through American self-reliance. Of course, producing all our own oil and gas will not protect us any better from storms, or terrorists, or from our own aging grid. It may not even protect us much against price spikes. The United States is already a net-exporter of natural gas and oil, but both are global commodities, and prices are necessarily a function of geopolitical dynamics and the market intervention of foreign governments. Starting in earnest with Hurricane Katrina in 2012, our national conversation around energy security has shifted from national energy independence to more targeted energy resilience.

We define energy resilience as the relative ability of an institution to continue its mission despite shocks to the energy system, whether they be economic, infrastructural, or climatological. Although the term *resilience* is sometimes used interchangeably with *reliability* in the energy context, the distinction between the two terms is important. The definition of energy resilience given above applies to all three of the "actors" in our analysis: grid, government, and local consumers. The term *reliability* refers to the provision of a service. For an electrical utility, the reliability of electric service is the desired outcome of the utility's resilience efforts. Reliability is the goal for governments and institutional consumers as well – they just provide services other than energy itself. A government might pursue energy resilience to ensure reliable provision of essential services like water, firefighting, and police protection. For a hospital, the ability to keep the power on during a utility outage is the non-negotiable cost of reliably providing life-saving services to its patients.

In contrast to the binary of energy independence, we use the concept of energy resilience to describe a spectrum. Each institution (whether utility, government, or consumer) occupies a particular spot on the energy resilience spectrum, based on that organization's ability to maintain continued operations during a disruption to the energy system. Importantly, it need not be the goal of every institution to push themselves to the hyper-resilient extremity of the spectrum. Leadership must determine the right place for their institution on the spectrum based on their specific risk profile and the resources they have to invest in resilience projects. As we will describe in this book, there are energy resilience tactics that can be pursued cost effectively for just about any institution. However, at a certain point, the marginal return on investment, either financial or in terms of actual resilience, may not justify additional action.

The Three Layers of Energy Resilience

Even as the concept of energy resilience has gained steam among policy makers and institutional managers, there are few examples of coordinated efforts among the various actors in the energy system. We argue that in order to affect the kind of systematic energy resilience necessary to meet the many risks currently faced by our energy system, grid operators, government regulators, and local consumers must engage in complimentary and concerted action. Throughout the book, we will describe how the energy resilience strategies and tactics employed by one of these actors naturally yield benefits for the other two, and how proper coordination can make these efforts more effective and efficient for all.

When interviewing her for this book, Marissa Aho, the Policy Director & Chief Resilience Officer for Houston, Texas told us: "I look at resilience by scale. If you do not improve resilience at each scale, none of the scales are resilient. People need to be resilient. Private businesses need to be resilient. Governing bodies need to be resilient too, but we can't be bound by our municipal boundaries, because they don't matter to nature. That's where regional partnerships are important. You need to pursue resilience at every scale as a coordinated effort."[4]

In this book, we intentionally restrict our analysis to the United States. Of course, the transformation of the energy system and the threats that it faces are, for the most part, universal, and many of the strategies and tactics we will discuss could apply just as easily to any other country. However, in order to more thoroughly examine the particularities of the American electrical grid and the government structures that regulate it, we found that it was useful to limit the scope of the book to the borders of our home country.

Grid Resilience

In addition to mounting external threats, grid operators are now contending with fundamental shifts in the energy system itself. As solar and wind generation have fallen in price over the past decade, their increasing presence on the grid has forced utilities to develop strategies for adapting to their intermittent generation profile. At the same time, there is a movement underway to "electrify everything" in the building and transportation sectors to minimize point-of-use emissions. The demand for electric vehicles and heat pumps is growing quickly, promising to significantly expand the loads that electric utilities must serve. This is both a business opportunity and a challenge for utilities, who must match this expanding load to an increasingly intermittent generation portfolio. Finally, the proliferation of distributed energy resources (both generation and storage) is forcing utilities to reimagine and redesign the grid to support new transmission and distribution requirements. As we will discuss, there are many solutions already on the market to address these challenges, but they will require upgrades to both physical and digital infrastructure to create a grid that is automated, flexible, and responsive to changing conditions.

Local Resilience

As much as climate change is a global process, its effects on people's lives are local. The same is true of our energy system. The effects of a grid disruptions can turn quickly from inconvenience into danger for communities and individuals, as heating or cooling needs go unmet, food spoils, and medical equipment shuts down. For the purposes of this book, we define the local actor as any institution that occupies a defined, contiguous space, and has control over the infrastructure on that space. This definition includes the government of a town or small city or the administration of a local institution like a hospital, university, or business. All of the institutions we discuss have in their mission a duty to protect the health and safety of their constituents, whether residents, patients, students, employees, or customers. Throughout the book, we will ground our analysis in the connection between energy security and human wellbeing.

Government Resilience

More than nearly any other industry, the history of the energy sector in the United States is intertwined with government regulation. The regulatory model has changed through the years, with the sanctioning of independent generators in the 1970s and utility restructuring in the 1990s. The changes in the energy system that are now underway may very well represent the biggest challenge energy regulators have ever had to address. To meet their mandate to ensure reliability, support the industry, and protect customers, elected officials and regulators will need to rapidly modernize regulatory frameworks to fit a more distributed, intermittent, and expanding grid, and do so in the face of growing external threats. Regulation is not the only role government has to play in creating a more resilient energy system. Federal and state governments can support this transition through research, grants, tax incentives, and the power of their own purchasing.

The Organization of this Book

This book is divided into three main parts:

Part 1: Energy Resilience Principles

Part 1 of the book is intended to give context to the energy resilience discussion and establish a conceptual framework for our analysis. In Chapter 1, we describe the many and growing threats to our energy system. We present the evidence for each and describe how it has impacted our three types of actors. In Chapter 2, we discuss the history of the energy resilience framework and how we apply it to each of these three types. We also describe the Spectrum of Energy Resilience that we employ throughout the book.

Part 2: Energy Resilience Strategies and Tactics

The second part of the book shifts away from the conceptual discussion presented in Part 1 and toward more concrete guidance for pursuing energy resilience. Chapter 3 describes the energy resilience planning process. Synthesizing lessons from the many planning frameworks that have been developed over the past five years, we discuss the steps involved in bringing together the relevant parties, identifying goals, and pursuing those goals in a manner that is most likely to achieve success. Chapter 4 presents a "maturity model" designed to allow managers to assess where their institution lies on the energy resilience spectrum and plot a course toward where they would like to be.

Part 3: Energy Resilience at Every Level

In the final section of the book, we dive deeper into the energy resilience imperatives and opportunities of each of our three main actors: grid operators, the government, and local consumers. In Chapters 5–7, we discuss how the threats we established in Part 1 affect each group differently and identify the best strategies for each in advancing energy resilience. We also discuss how energy resilience projects pursued by one actor may be complimentary to those pursued by others, and how the combined efforts of all three can total more than the sum of their parts.

Case Studies

The conclusions in this book draw on the authors' combined experience in the energy industry, working as consultants and technical solution providers to governments, local institutions, businesses, and utilities. In researching this book, we also conducted extensive interviews with leaders of all of these organization types, including city energy and resilience managers, federal government energy experts, and utility executives. We have developed case studies based on many of these conversations and have deployed them throughout the book to better illustrate the real-world challenges these organizations face, and the successes they have achieved. In doing so, we hope that the book will provide practical lessons to those tasked with advancing energy resilience in their own organizations and spur greater action and collaboration to that end.

Part 1

Energy Resilience Principles

1

The Energy Resilience Imperative

The dependable access to energy we enjoy in the United States is based on a complex relationship of natural, technological, and institutional factors. Each flip of the light switch relies on machines and institutions adequate to extract fuel from the natural environment, transform it into electricity, and distribute it over tens or even hundreds of miles. Every step in this process, in turn, relies upon social organization and support, from international markets for fuels and electricity to the public/private coordination that supports the generation and distribution of energy on a large scale. For such a complex process, it is remarkably reliable. However, when just one of the necessary natural, technological, or institutional conditions fails, we quickly find ourselves in the dark.

In this book, we discuss energy resilience as the ability of institutions and communities to continue to safely operate and pursue their mission amid threats to the energy system. The path to energy resilience will be different for a hospital than for a small town or for the grid itself, but leaders in each should be aware of the factors that pose potential threats to energy security and be able to assess each threat to ensure that it is properly addressed. Long-term fuel supplies and price volatility both present risk to energy security but are often eclipsed by aging energy infrastructure, severe weather events, and sabotage. Looming over all these is the changing climate, which is exacerbating existing weaknesses in our energy system, and carving out new ones.

In this chapter, we will discuss the nature and severity of threats to the natural, social, and technological systems on which the U.S. energy system relies. First, we will discuss a few of the most important examples of these prerequisites: access to fossil fuels, the water/energy nexus, infrastructure, and sabotage. We will finish by discussing how climate change runs through and exacerbates all of these vulnerabilities.

Access to Fossil Fuels

The presumption of centralized, fossil fueled bulk power generation is baked into every aspect of our energy system, from grid regulation to transmission and distribution design to rate structure. The addition of new clean and distributed generation to the grid has therefore been a challenge. There is a school of popular thought in the industry that natural gas will provide reliability going forward, without requiring significant redesigns of the grid or its governance. However, even access to natural gas faces its own growing challenges – some natural, some social – and an energy system that is based on natural gas generation may prove less resilient than we hoped for in the face of climate change. The same may be said for oil to power our transportation. If we instead electrify road travel and replace natural gas with renewable sources, we will face different reliability and resilience challenges. We will discuss these briefly later in the chapter and in more depth in Chapter 5, "Grid Resilience."

For decades, we have been struggling to make the transition from a world in which the supply of fossil fuels seemed endless, to one in which the limits on that supply are making the extraction of these resources increasingly more difficult and expensive. As we have approached the limits of existing extraction capacity, we have sought new, more expensive, or technologically complex means of gathering fuel. To find untapped reserves, we have moved deeper into the ocean and further into the arctic than previously thought possible. We have drilled in formerly protected lands and devised technology to extract gas from shale formations that were previously unreachable. In doing so, we have managed to keep up with ever-rising demand. However, our own ingenuity will only hold out so long against a finite resource. If widespread adoption of non-fossil energy sources does not occur, the gap between effective supply and demand for fossil fuel will widen and new stresses will arise on our energy security, from price volatility to social conflict and climate change.

Over the past 20 years, hydraulic fracking (the process of using pressurized liquids to fracture bedrock formations in oil or gas wells) has both lived up to the promise of its promoters, and the warnings of its detractors. For the past 10 years, hydraulically fracked wells have made up a majority of new wells drilled, and the combination of horizontal drilling and hydraulic fracturing has stimulated crude oil and natural gas production in the United States, both of which reached record levels before the COVID-19 pandemic in 2020.[1] For the rural areas above major shale formations, the development of new fracking wells has meant financial opportunity for land-owners and the potential for a stimulated local economy.

These same communities, however, bear the brunt of the environmental impacts that accompany fracking. In a report to Congress, the Government Accountability Office found that "shale oil and gas development poses risks to air quality, generally as the result of engine exhaust from increased truck traffic, emissions from diesel-powered pumps used to power equipment, gas that is flared (burned) or vented (released directly into the atmosphere) for operational reasons, and unintentional emissions of pollutants from faulty equipment."[2] At the same time, EPA found a variety of ways in which fracking has the potential to significantly impact drinking water in these communities, from withdrawals of high volumes of fresh water for fracking to the accidental or intentional contamination of groundwater and surface water with fracking liquid and insufficiently treated wastewater.[3] In December 2020, Physicians for Social Responsibility and Concerned Health Professionals of New York released a thorough meta-analysis of research on the effects of fracking on economic, environmental, and public health. On the latter point, they found that "Public health harms now linked with drilling, fracking, and associated infrastructure include cancers, asthma, respiratory distress, rashes, heart problems, and mental health problems. Multiple studies of pregnant women living near fracking operations across the nation show impairments to infant health, including birth defects, preterm birth, and low birth weight."[4] As with most environmental impacts, the greatest harm is experienced by the most vulnerable communities – in particular poor, rural farming communities that have played host to the majority of new fracking well development.[5] The ill effects of fracking were sufficient to lead four states to ban the practice, including New York, which sits upon a potentially oil- and gas-rich swath of the Marcellus shale formation. In April 2021, Governor Gavin Newsome announced his own intention to ban the practice in California by 2024, to be followed by all other forms of oil extraction by 2045.[6]

California is often a policy outlier in energy and other areas, but if fracking is made increasingly more difficult due to regulation, it decreases the potential of natural gas to provide reliability and resilience into the future. Increased barriers to domestic fracking may lead to more price volatility and the need to venture further and build longer pipelines to deliver the fuels to their point of use.

In addition to fracking, arctic drilling was seen as an important new frontier in domestic oil and gas exploration. Over the past 10 years however, efforts to develop resources in the arctic have proven much more challenging than terrestrial fracking, and just as unpopular. It was hoped that climate change may actually open up new channels for oil exploration, as rising temperatures allow ships and rigs to reach areas previously blocked by sea ice. However, disappearing ice may actually make it prohibitively difficult to set up the kind of infrastructure needed to support new drilling operations. In 2018, the Hilcorp Liberty project was slated to be the first new oil production project in federal arctic waters after the Trump administration lifted an Obama-era ban on such projects. However, shortened arctic winters slowed construction, as the ice formations necessary for accessing the project over land were forming later and melting earlier than ever before.[7] In 2020, the project was halted completely when a federal appeals court found that associated environmental review had been insufficient. Hilcorp's story is not an outlier among similar efforts in the arctic. Shell Oil was one of the first movers in the area, sending rigs north starting in 2012. Early efforts were a disaster for Shell, with equipment running aground, failing, and even catching fire. The Obama administration gave Shell a second chance in 2015, but Shell completely abandoned plans for arctic exploration later that year, after investing nearly $7 billion in the effort.[8] The lessons of these recent efforts seem to have left an impression on the rest of the industry. When drilling rights in the Arctic National Wildlife Refuge finally went on sale in January 2021, no major oil companies bid, and only half of the leases offered for sale garnered any bids at all. This was a blow to the outgoing Trump administration, which had projected that opening the refuge to oil exploration could yield a billion dollars in lease agreements. Instead, the sale brought in less than $15 million.[9]

Even if fossil fuels are successfully extracted from the ground, they must still be delivered from remote or rural areas to where they will be processed and used. The United States' natural gas infrastructure now includes 318,000 miles of transmission pipelines and 2.28 million miles of distribution pipelines. Most of this pipeline is owned and maintained by one of hundreds of private entities. In recent testimony before congress, the Director of Homeland Security William Russell reported that, "Interstate pipelines run through remote areas and highly populated urban areas, and are vulnerable to accidents, operating errors, and malicious physical and cyber-based attack or intrusion. Pipeline system disruptions could result in commodity price increases or widespread energy shortages."[10] Like the extraction practices discussed earlier, the construction of pipelines has also proved to be controversial and has elicited vigorous and sustained protest over environmental damage and eminent domain. There can be little doubt that this activism played a role in halting two of the most prominent recent projects, the Keystone and Dakota Access pipelines – projects which, for activists, came to embody the fight over issues from tribal sovereignty to the conversation of public lands and the protection of critical waterways.

When you consider the documented leakage of natural gas from extraction facilities and pipelines, domestic natural gas starts to seem even less like a panacea for American energy security. Natural gas emits less carbon pollution per unit of energy produced than coal or other petroleum products. Methane itself, however, is a potent greenhouse gas – some 84 times more potent than CO_2 over a 20-year period. Recent studies have used the chemical signature of methane to confirm that leaks from natural gas infrastructure are the cause of rising concentrations of the gas in the atmosphere.[11]

The sum of the logistical difficulty, environmental damage, and social antipathy associated with the extraction and transportation of fossil fuels mean that new exploration for gas and oil is unlikely to provide the solution to growing energy resilience challenges.

As we seek to build a more resilient energy system, we need to make choices about what that system will look like, and what models we should use to guide us. Continued reliance on natural gas and oil may seem like the easy answer. Our energy system is already designed for bulk power generation based on easily deployable fossil fuels, our highways are dotted with gas stations, and our homes are built with natural gas in mind. However, it is important to face the growing challenges and cost associated with delivering these fuels before we make investments that double down on this model. Even if the relationship between burning fossil fuels and the risks posed to the energy system by climate change are set aside, investments in fossil fuel infrastructure are still not likely to yield the greatest resilience benefit when compared to alternatives.

The Energy Water Nexus

Supply of fuel is not the only variable in the formula that brings energy to our homes and business. Increasingly, scientists are pointing to water as a possible limiting factor in this equation.[12] Water is a key ingredient throughout the energy supply chain. It is used extensively in the drilling and mining of fuels, including natural gas, coal, oil, and uranium. Fuel refining and processing is a water-intensive process in its own right and is required before oil, uranium, or natural gas can be used in a power plant. Transporting the refined fuel uses still more water. Water is used to test fuel pipelines for leaks and to transport coal through slurries. Finally, water is essential to the function of a vast majority of the power plants in the United States.[13]

About 90% of power generation in the United States is thermoelectric – a term used to describe the process of using heat to create steam and turn a turbine that generates electricity.[14] Thermoelectric power plants generate heat through the combustion of fossil fuel, the fission of nuclear fuel, or the concentration of solar radiation. Water is used to cool the steam that turns the turbines in these plants and sometimes to condense it for reuse. More water is withdrawn from natural reserves for thermoelectric power generation in the United States than for any other purpose: over 40% of the total.[15] In addition to withdrawing

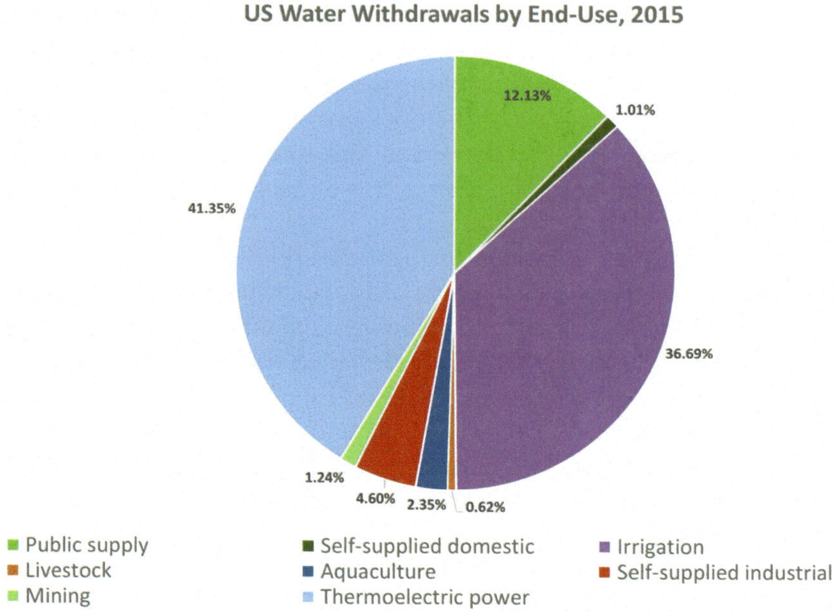

Figure 1.1 Data from "Estimated use of water in the United States in 2015." Dieter et. al. USGS. Circular 1441. 2018.

more freshwater than any other industry, energy generation is also one of the largest non-agricultural *consumers* of freshwater. Water consumption refers to the portion of withdrawn water that is not available for reuse because it is lost to evaporation during the heating and cooling process in power plants or because it becomes polluted and must be treated as wastewater.

This great reliance on freshwater makes the energy supply chain, and power plants, in particular, vulnerable to variations in the water supply. Indeed, we are already seeing cases of powerful energy companies losing the battle over scarce water resources. In a report published by the Union of Concerned Scientists, "Water-Smart Power: Strengthening the U.S. electricity System in a Warming World," Rogers *et al.* note many instances between 2006 and 2012 of plants that had to reduce their output or shut down completely because available water supplies were insufficient (either in volume or temperature) to run the plant effectively.[16]

The power and water disruptions that followed the 2021 winter storms in Texas provide a particularly daunting example of the interplay between the two systems. The state's grid was unprepared for level of cold brought by the unusual (though not entirely unprecedented) far reach of the so-called polar vortex, and disruption to the power system was severe. Coal stores froze, preventing the fuel from being moved into power plants. Some wind turbines that had not been winterized also froze and were taken offline. Most importantly, many of the state's natural gas plants were shut down by cold temperatures at the plant, or after natural gas pipelines were blocked by frozen water (there is water vapor in natural gas pipelines), starving the plants of fuel. In the end, five times more power was lost from natural gas generation than from wind.[17]

This initial damage triggered several positive feedback loops that exacerbated the impact on the power system and on consumers. When natural gas plant shutdowns started to trigger outages, the compressors that are needed to move natural gas through even unblocked pipelines themselves lost power, compounding the difficulty of delivering the fuel.[18] Cold temperatures caused demand for natural gas for heating homes and buildings to peak, adding to overall shortages and sending prices soaring. Some natural gas plant operators, unable to turn a profit with fuel prices so high, elected to instead take their plants offline.[19]

And then there was water. Frozen water in fuel pipes caused electricity and natural gas outages for consumers. Water pipes froze in mains, homes, and buildings, causing flooding and dangerous shortages of potable water. Power outages at water treatment plants exacerbated these shortages. Finally, system impacts prevented water from getting to power plants, where it was desperately needed to get the thermoelectric system up and running again. Nine days after the first storm struck Texas, there were disruptions in more than 1180 public water systems across 160 counties, leaving 14.6 million people without water, many of them still sitting in the cold and dark.[20]

In Nevada and California, climate change and the energy/water nexus are affecting electric power generation in another way. Hydroelectric power plants are not thermoelectric. Instead of using steam to generate power, they rely on running liquid water to turn turbines directly. Water scarcity, therefore, has a rather obvious potential to limit the output of hydroelectric plants, and we have seen this play out during recent severe droughts in the western United States. A 1% reduction in the flow of the Colorado river, for example, can reduce energy output from the river's various hydroelectric plants by 3%.[21] In California, cheap and clean hydroelectric power typically accounts for about 15% of electricity consumed in a non-drought year. 2021 was one of the driest years on record in California, and, overall, the drought that has spanned the years 2000–2021 has been the worst in at least 1200 years. The dry conditions have taken a severe toll on the state's hydroelectric generation, which now accounts for just 7% of total power produced.[22] At the time of this writing, water levels in the Lake Oroville reservoir are so low that operators expect to have to close the Edward Hyatt Power Plant that it feeds – something they have never had to do in the 54 years of the plant's operation. Capacity at the plant has already been reduced to 80%. At Lake Mead in Nevada, water levels are so low that the famous Hoover Dam is down to 75% of total capacity.[23]

As we will discuss in more detail later in the chapter, all of these pressures are amplified by the mounting effects of climate change. Over the past decade, we have seen droughts become longer, more frequent, and much more severe. Even under normal precipitation conditions, higher average temperatures encourage evapotranspiration and lead to more irrigation for agriculture, creating a greater need for freshwater in the industry that is thermoelectric's greatest competitor for the resource.[24] Of course, it is not only water that is in greater demand in warm conditions, and space cooling needs also spike during these times. As temperatures rise, energy producers will need to find ways to produce more energy with less cool water.

We are in a time of rapid transition in the energy sector, with inexpensive natural gas increasingly taking the place of coal power, and wind and solar (which require no water for steam generation or cooling) making quick gains through market-beating prices. Since the lives of power plants are generally measured in decades, the technological and policy choices we make for each new plant will have significant implications for the future of our energy system. Rogers *et al.* write: "Understanding and addressing the water impact of our electricity choices is urgent business. Because most power sector decisions are long-lived, what we do in the near term commits us to risks or resiliencies for decades. We can untangle the production of electricity from the water supply, and we can build an electricity system that produces no carbon emissions. But we cannot wait, nor do either in isolation, without compromising both."[25]

Electricity Infrastructure

Periodically, the American Society of Civil Engineers (ASCE) releases a report card assessing the strength of American energy infrastructure, including evaluations of overall capacity, energy grids, other forms of distribution, and investment for the future. After earning scores of "D" or "D+" on every report since the first one in 1988, energy infrastructure made a breakthrough in the 2021 report, pulling in a "C–."[26] ASCE elaborates that over the past four years, investment in energy infrastructure has indeed risen somewhat as utilities have sought to harden their infrastructure. The organization nevertheless concludes that infrastructure investment is woefully insufficient to meet the needs of an energy system that is quickly transforming, and facing an onslaught of wildfire, drought, and other severe weather.

Our energy infrastructure was not developed as a cohesive national project, and, today, it is an amalgam of generation, transmission, and distribution equipment, strung together across regions as they have come online. This includes approximately 5800 major power plants, a network of over 450,000 miles of high-voltage transmission lines, and all of the local overhead lines or underground cables that bring power to each home and building.[27] Most of this infrastructure was developed decades ago, and some dates back to the 1880s. The ASCE has cited the age of this infrastructure as a primary factor threatening its reliability. The authors of the 2021 Report Card found that much of our nation's energy infrastructure is operating past its intended life span, and that 70% of transmission and distribution lines are "well into the second half of their lifespans."

Age alone does not account for the increasing fragility of our national energy system. This infrastructure is being called upon to perform functions today that engineers could scarcely have imagined when much of it was created: increasing use of air conditioning, electric vehicle charging, and the high load, uptime, and power quality demands of facilities like data centers, to name a few. As our use of energy and electricity has evolved, infrastructure has lagged behind.

For most of its history, the electric power grid was owned and operated by a network of vertically integrated utilities, each providing both generation and distribution services within their geographic territory. Each utility operated as a regulated monopoly. There was little competitive pressure, and regulators ensured that the utilities were delivering power reliably, with minimal risk to the system. In the 1990s, roughly half the states deregulated their energy markets (also known as utility restructuring). Under these regimes, states required utilities to divest from their generation assets, and created an open market for electricity supply. Utilities faced some reduced regulation of their planning and investments but now had to source power generation from third parties. The result is what University of Minnesota Professor Massoud Amin referred to as a less "shock absorbent" system.[28] Utilities no longer had full control over the makeup and load profile of all available generation sources. Amin writes, "as a result of these 'diminished shock absorbers,' the network is becoming increasingly stressed, and whether the carrying capacity or safety margin will exist to support anticipated demand is in question."[29]

In the decade since Dr. Amin made those observations, systemic stresses on the grid have grown. The most significant of these is severe weather, which we will discuss later in the chapter. Indirectly, climate change is creating other systemic changes to the ways we generate and use electric power, and challenging the existing grid's ability to adapt. First, there is a growing belief among activists, policy makers, and even industry that in order to address the climate crisis, we must "electrify everything." This means that in parallel with cleaning up bulk power generation, we must replace natural gas and gasoline as fuels to heat our homes and businesses, cook our food, and power our cars and lawnmowers. Transportation alone accounts for 29% of GHG emissions in the United States, and residential and commercial buildings add another 12.8%.[30] If all the power for our cars, trucks, and buildings could come from electricity, their emissions profile would drop along with grid itself as clean energy replaces fossil fuels in power plants. However, this outcome would also create an enormous increase in electrical demand. A 2018 study from the National Renewable Energy Laboratory (NREL) found that a scenario associated with high levels of electrification of end-uses would be associated with a nearly 40% increase in electricity consumption by 2050.[31]

Inexpensive clean power sources such as wind and solar are likely to be a big part of meeting this increased demand, due to both market forces and public policies designed to stem emissions. However, wind and solar add yet another challenge – the intermittency of their production. Instead of providing consistent baseload power that can be throttled on demand like a gas peaker plant, wind turbines and solar panels only generate power when the wind is blowing and the sun is shining, respectively. Utilities will need to upgrade and redesign the grid to adapt to both growing demand and increased intermittency. This will mean increased grid segmentation and automation to allow for better geographic balancing of load with supply. It will also mean the adoption of grid-scale energy storage to shift electricity supply temporally from times of peak generation to times of peak demand.

Certain resilience projects are clear infrastructure projects (such as undergrounding lines or hardening of above ground lines) and can, therefore, be rate-based. This means that utilities can recover the cost (plus profit) of these projects by adjusting rates they collect from customers. Capital investments in the grid are not the only solution, however – and often not the most cost effective. Solutions like energy efficiency, demand response, and third party distributed energy projects such as microgrids are often referred to as "non-wires alternatives." Such projects can mitigate peak demand and balance the local grid but may require utilities to forego capital projects that they could profit from, and instead pursue projects that yield no profit. In traditional "cost of service" regulatory models, there is little incentive for utilities to make these choices. The "Grid Resilience" chapter of this book goes in depth on ways in which utilities can work with their regulatory agencies to harmonize economic incentives with grid investment needs.

Case Study – Hurricane Resilience in New Orleans

New Orleans is one of the areas in the United States most vulnerable to climate change, given population density, elevation at or below sea level, and its location directly in the path of many of the worst hurricanes to make landfall in North America. This vulnerability was demonstrated for the entire world by the devastation that followed Hurricane Katrina – an event that spurred modern efforts to make our energy system more resilient to climate change. The city has been hit with a series of severe whether events since Katrina, and although none wrought the devastation of that 2005 storm, it is clear that energy resilience in the city is still a work in progress. In 2017, the city's sole electric utility, Entergy, proposed a new 128-megawatt natural gas power plant, to be located within the city in a predominantly Black and Vietnamese area. The project met vocal resistance within the community, and opposition grew following the revelation that many who showed up to city council meetings to support the project were paid by a firm that the utility had hired. Entergy argued that the plant was necessary to provide a source of energy inside New Orleans should the eight transmission lines that feed the city be knocked out in a storm.[32] The utility promised that the plant had "black-start" capability that would allow it to kick-on quickly after power was lost from other sources. The plant was built, and in 2021, the city was forced to test Entergy's premise.

Hurricane Ida struck on August 30, knocking out all eight of the transmission lines that feed New Orleans, and leaving over 1 million customers without power. The black start function of Entergy's new gas plant seemingly failed, and it took two days for the plant to restart following the storm. The plant was only ever designed to serve certain critical and vulnerable loads; so even if it had operated as designed, many homes and businesses would still have been left without power.[33] The cause of the broader problem seems to have been damage to lines and poles. In addition to transmission lines, over 31,000 distribution poles were also damaged in the storm. Energy has said that all its poles were designed to withstand winds according to requirements when they were installed. The problem is that these requirements have changed over time. Hurricane Ida was a Category 4 storm, and knocked down power lines with wind gusts up to 150 miles per hour. The New York Times reported on documents from McCullough Research that found that much of Entergy's infrastructure in and around New Orleans was built to withstand winds only up to 110 miles per hour, or a Category 2 storm.

As Category 4 storms become more common along the Gulf Coast, it is clear that the city of New Orleans, the State of Louisiana and electrical utilities need to protect all their infrastructure against such storms. The special relationship between utilities and regulators makes it difficult to assign all blame to one entity or another. The utility is regulated by the City of New Orleans inside its boarders, and by the Louisiana Public Service Commission in the rest of the state. Broadly speaking, it is the responsibility of these regulators to approve projects and associated rate increases to improve resilience and to require certain levels of resilience from the utility. The story in New Orleans is going to be common to just about every corner of the country as we strive to upgrade grid infrastructure in the face of climate change. Regulators need to make this a priority, utilities need to invest wisely and be honest about progress, and public investment may be required to make the progress we need while still controlling energy costs for vulnerable customers.

At the national level, we will need a better integrated, more responsive grid. Non-wires alternatives will help to alleviate the burden of intermittency, but these will not replace the need to build a grid that can nimbly direct bulk power when and where it is needed. Strategic investments in transmission capacity will broaden the areas across which we can balance supply and demand. A recent NREL study determined that connecting the three major components of the U.S. power system (the Western Interconnection, the Eastern Interconnection, and the Electric Reliability Council of Texas) to allow for electricity flow between geographies would have a benefit cost ratio of 2.9, indicating a huge value to the economy from greater transmission interconnection.[34]

Whether a failure occurs in the delivery of fuel or the distribution of electricity, it is likely that the effects will be felt broadly within our energy system. Our energy infrastructure is highly interconnected and interdependent and was not designed to promote the resilience of the entire system when one part fails.

Terrorism and Sabotage

Only a few generations ago, an energy bill was known as a "light bill" because, at least in households, electricity was used for little else. Today, nearly every function of society relies on a constant, reliable stream of energy. From banking to agriculture and transportation to government, much of what we do shuts down during a power outage, with effects of local outages rippling out through sectors and across the country. Any failure in our energy infrastructure can turn into a threat to our security, productivity, and economic welfare. This dependency, and the damage that could result from a large-scale power disruption, makes the energy system a particularly attractive target for saboteurs. Potentially adding to this allure is the vast interdependency of the energy system itself, and the fact that a successful attack on a certain piece of the system is likely to cause cascading failures.

Seeking to better understand this vulnerability, the Department of Homeland Security asked the National Research Council (NRC) to study to matter. In 2005, the NRC convened the "Committee on Enhancing the Robustness and Resilience of Future Electrical Transmission and Distribution in the United States to Terrorist Attack," to assess the threats, develop recommendations for mitigating them, and issue a report to the government. This report was completed in 2007 but, due to security issues, was not released to the public until 2012.[35] The NRC researchers describe an energy system in the United States marked by significant security exposures, falling into three general categories: physical, cyber, and personnel vulnerabilities.

Even as physical energy infrastructure has aged, digital monitoring, communication, and automation systems have been built on top of it. Computer systems are now used throughout the grid to control equipment, analyze reliability, and manage the energy market. These are all potentially susceptible to sabotage, but the NRC report singled out supervisory control and data acquisition (SCADA) systems as the most vulnerable. These systems monitor substations and send control signals to other equipment, such as circuit breakers, to keep transmission and distribution systems in balance. Unlike damage to other computer systems, which may stay contained, an attack on a SCADA system could have the potential to cause widespread outages, as entire transmission or distribution networks could be taken offline.[36]

The existence of a network that is not strictly local gives saboteurs an opportunity to affect these systems remotely, including from another country. In June 2014, several cyber security firms discovered that a hacker group originating in Russia had been aggressively infiltrating the control systems of oil, gas and power companies in the United States and Europe. This group, known alternately as "The Energetic Bear" and "Dragonfly," used trojan-horse hacks to gain access to the industrial control software used to manage the grid. It is assumed that the intent of these attacks was industrial espionage, but if they were so inclined, the control wrested by the hackers could certainly have been used to cause major disruptions to power supplies.[37]

Incidence of cyberattacks on energy infrastructure has grown more frequent and more sophisticated since that 2014 "Energetic Bear" incident. In addition to political motives or a desire to sew chaos, some attackers are now motivated by monetary gain. A cyberattack on the Colonial Pipeline in 2021 illustrated the potential impact of adding profit-motivated criminals in the list of potential adversaries to our energy system.

The Colonial Pipeline attack used a lever commonly referred to as "ransomware," in which a target is disabled, and the attacker makes it clear that their systems will only be put back online once a ransom has been paid. In this case, the system in question was one of the largest petroleum pipelines in the United States, carrying almost half of the gasoline, jet fuel, and diesel consumed on the East Coast. For all those Americans who witnessed the effects of this attack on the news or in their own lives, it may come as a surprise to learn that the hackers never actually gained access to the pipeline operational computer systems. The attack was on Colonials administrative or business-side data systems, but the company was so concerned that the hackers may be able to expand their access into operational data systems that they preemptively shut their operations down, including all materials flowing through their pipeline. This set off a chain reaction of ever-worsening effects, from scarcity or perceived scarcity at east coast gas pumps to hoarding behavior, which in turn exacerbated scarcity and price gauging. By the time Colonial paid its extortionists nearly $5 million in ransom, the U.S. Energy and Homeland Security departments had determined that the country could only withstand another three to five days of shuttered pipeline before public transit systems would have to curtail operations, and petroleum refineries would have to close down due to lack of an outlet for their product.[38]

Following the Colonial Pipeline attack, Fatih Birol, the head of the International Energy Agency tweeted, "The ransomware attack on the #ColonialPipeline in the U.S. shows the critical importance of cyber resilience in efforts to ensure secure energy supplies. This is becoming ever more urgent as the role of digital technologies in our energy systems increases." Unfortunately, it is not easy to manage the resilience of our energy infrastructure to cyberattack, since so much of this infrastructure is owned and managed by private companies, each with their own approach to cyber security. The Biden Administration released an Executive Order in 2021 aimed at expanding the kind of public/private partnership and coordination that will be needed to bolster our defenses, but there is a tremendous amount of ground to cover, and our adversaries are learning right along with us with each new attack.

The NRC report concludes that although cyberattacks would not take nearly as long as a physical attack to repair, the threat is still a serious one, and a cyberattack could be combined with a physical attack to create a disruption larger than the sum of its parts. For instance, a cyberattack could be used to obfuscate a physical attack at a remote location, thus delaying the response.

Climate Change

The impacts of climate change are apparent all around us. What was once as subtle as a half degree rise in ocean temperatures or as distant as a thinning arctic ice shelf is now stark, and frighteningly immediate nearly everywhere humans live. It is not difficult for even those who otherwise spend little time thinking about the energy system to see the impacts of climate change on access to power. The most obvious manifestation of climate change may be severe storms, but changing environmental conditions present a variety of other threats to our energy security.

To examine these threats more closely, we will break them down into their proximate causes. Specifically, we will look at four main effects of climate change and their implications for our energy system:

1. Increasing air and water temperatures
2. Decreasing water availability
3. Increasing frequency and intensity of storm events
4. Sea level rise

These effects of climate change are the same as those identified in a 2013 report from the U.S. Department of Energy titled "U.S. Energy Sector Vulnerabilities to Climate Change and Extreme Weather."[39] Of these risks, severe weather events have received the most attention, and they do pose a serious risk. However, other climate-change-related factors (such as water scarcity, as we have discussed) will affect the energy system in a variety of ways – some more subtle than a hurricane but no less perilous.

Rising Temperatures

Perhaps the most obvious of these factors is increasing global temperatures. Average seasonal temperatures have been rising over the past century, and at an increasing rate. 2020 was the second hottest year on record and the top 10 hottest years ever have all occurred since 2005.[40] Rising overall temperatures have manifested as intense and more frequent heat waves, longer wildfire seasons, thawing and shrinking permafrost cover and sea ice, and a longer growing season.[41]

These conditions threaten the energy system in a variety of ways. In the arctic, temperatures are rising even faster than the global average, and the changing landscape is compromising oil and gas infrastructure. Oil and gas resources in Alaska are critical to the U.S. energy portfolio, but as permafrost thaws, it loses its ability to bear weight, imperiling pipelines, extraction equipment, and even buildings that have been erected to support the industry. The Alaska Department of Natural Resources limits the number of days per year that travel is allowed on the tundra to help protect its integrity, and over the past three decades, this number has fallen from 200 days to just 100 days, effectively cutting the amount of oil and gas activity that can occur throughout the year by half.[42] As we increase our reliance on northern territories for fossil fuel drilling, these environmental changes will make fuel more difficult, expensive, and dangerous to extract.

Rising water temperatures may limit the efficiency and effectiveness of power generation. Cool, fresh water is vital to every thermodynamic power plant, as it is needed to affect the phase change of steam back to water following the turbine. The pressure change between these two states drives the steam through the turbine, and any increase in water temperature decreases its ability to create this pressure differential. Increasing water temperatures also may make it difficult for certain power plants to meet regulations that limit the temperature of discharged water as means of protecting aquatic environments.[13]

Energy transmission may also be compromised by rising temperatures. The proportion of energy lost in transmission and distribution (currently about 7%) will increase as temperatures increase, and the carrying capacity of the lines themselves will decrease. Increasingly frequent and severe wildfires resulting from rising temperatures and drought are already destroying more remote energy distribution and transmission systems. These transmission systems will be strained further as rising temperatures increase the demand for energy used for cooling. We may see a decrease in fuel oil and natural gas used for heating during the winter, but peak demands in summer months are likely to increase.

The wildfires in California in 2019 illustrate how the various effects of rising temperatures can conspire to cause cascading energy crises. Warm air temperature encourages evapotranspiration, leaving the ground and flora dryer and more susceptible to fire. Heat can also cause power lines to sag and arc, throwing sparks that can ignite nearby vegetation. Another effect of climate change, drought, can turn a spark into a wildfire. In 2019, the fires were not as severe as that in the two previous years, but their effect on electrical power was much more profound.

California utilities determined that the only way to control the incidence of wildfires in particularly dry, windy conditions was to shut off lines that posed a particular risk. During the months of October and November 2019, utilities in the state intentionally shut off power to more than a million customers, representing approximately three million people, spread across 30 counties. PGE alone shut off power to 960,000 customers in one day, on October 26. These "Public Safety Power Shutoffs," or PSPS, can be their own kind of disaster. Many of the PSPS in 2019 were carried out with little notice, and in some cases, no notice at all. PSPS are likely to be a part of life in California for years to come. Following the wildfire season in 2019, all California utilities were required to submit wildfire mitigation plans to the California Public Utility Commission. All of the plans included continued PSPS.

Finally, air conditioning during heat waves are creating enormous load spikes for utilities, especially in urban areas. White Remer, in the *Sustainability & Resilience Officer for the City of Tampa*, described it this way: "I know hurricanes are an easy thing to go to here in FL, but the more I look at the research and listen to the community, extreme heat is rising in the ranks as an issue that could disrupt our activities and have major consequences for our community. Extreme heat is hardest on the most vulnerable. The Union of Concerned Scientists is telling us that we have a lot of extreme heat headed our way in the coming years. This is a silent killer. It's one of the major initiatives in our resilience roadmap."[43]

Decreasing Water Availability

Warmer temperatures speed up the rate of water evaporation from the surface of the earth and increase the capacity of the atmosphere to hold water. This seemingly simple change has widespread effects, disrupting regional water cycles and altering precipitation patterns.[44] Drought conditions have increased over the last 40 years, and 2021 saw a continuation of severe drought conditions throughout the Western United States, including what has become the worst drought in California's history. In 2020 and 2021, less than half of the average precipitation fell over the western states. The summer of 2021 put over 88% of California in "extreme" or "exceptional" drought, according to the U.S. Drought Monitor.[45] Other effects of water cycle disruption include rises in rainfall in severe storms, more rain and less snow in cold climates, and earlier spring melts.

As we discussed earlier in the chapter, a decrease in the availability of fresh water may have serious consequences for our energy system. We will not rehash these effects here, but it is important to note that as climate change raises global temperatures, the energy sector is likely to find itself in greater competition for the water it needs to support its supply chain. Warm temperatures lead to increased demand for water for many of the processes we rely on for physical and economic health. Irrigation, livestock, and drinking reservoirs all require more water in warmer weather and are likely to find their supply less reliable as temperatures rise.

Increasing Frequency and Intensity of Storm Events

The disruption in water cycles we have just described can lead to more frequent and more severe storms. As temperatures increase, the atmosphere can hold more water. Increased water in the atmosphere causes precipitation events to become more intense and also affects the geographic distribution of rain and snow. As some areas are flooded, others are left in drought.[40] These extreme weather patterns pose a wide range of threats to our energy system at every stage of the energy supply chain.

Much of the equipment used in the United States to identify new fossil fuel reserves, extract those fuels, refine and process them, and transport them for sale are located on or near a coast or offshore. The Gulf Coast is central to the American energy system, and it is also particularly vulnerable to severe weather. Around 50% of U.S. crude oil and natural gas production occurs in the Gulf Coast area, along with nearly 50% of its refining capacity. Energy infrastructure there is so dense that Hurricanes Katrina and Rita disabled or destroyed more than 100 platforms, damaged 558 pipelines, and forced many refineries to shut down for a period of several weeks.[46] Although hurricane Sandy struck further north, it too did damage to our fuel production infrastructure, with nearly 7% of U.S. refining capacity in the direct path of the storm.[47] The Buckeye and Colonial oil pipelines were forced to close following the storm, accounting for 825,000 barrels of lost delivery per day. Sandy also halted shipments of fuel into New York Harbor, cutting delivery of fuel by nearly 60% in the days following the storm.

Energy transmission and distribution equipment is particularly vulnerable to severe weather. The number of incidents of grid disruptions from weather events appears to be on the rise (see Figure 1.1). Many of these

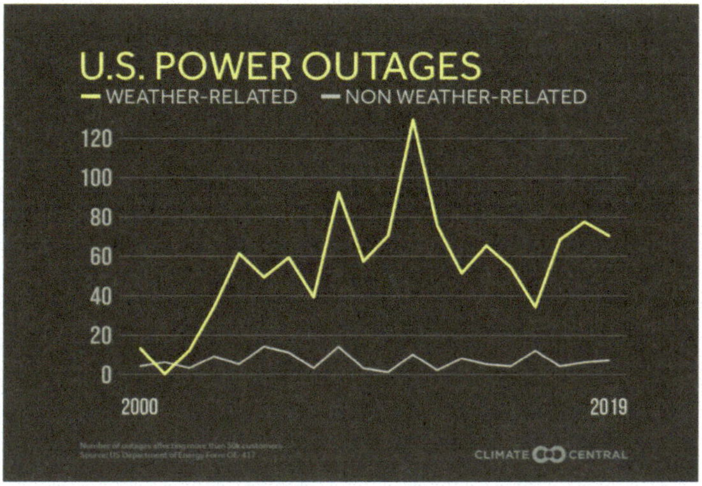

Figure 1.2 Climate Central. *Power OFF: Extreme Weather and Power Outages*. Sept. 30, 2020. ClimateCentral.org.

power outages result from trees falling on one of the thousands of miles of distribution lines strung around the country. In more severe cases, the effects of a storm can cascade across our fragile and interdependent grid, leaving large regions in the dark. Hurricane Sandy affected 8.7 million power customers, 1.4 million of which were still without power six days after the storm.

Sea Level Rise

The average global sea level has been rising at an increasing rate in recent years, following two millennia of little change at all.[40] As we have noted, much of our country's key energy infrastructure is located near the coastline, in low-lying areas inland, or offshore. As the forces of climate change cause sea levels to rise, this infrastructure will be made increasingly vulnerable to flooding from heavy rains and storm surge. Increasingly severe storms in coastal areas have the potential to damage not just energy infrastructure, but seawalls, levies, and other infrastructure designed to hold back storm surge and rising sea levels, further exacerbating the risk to any energy infrastructure that becomes exposed.

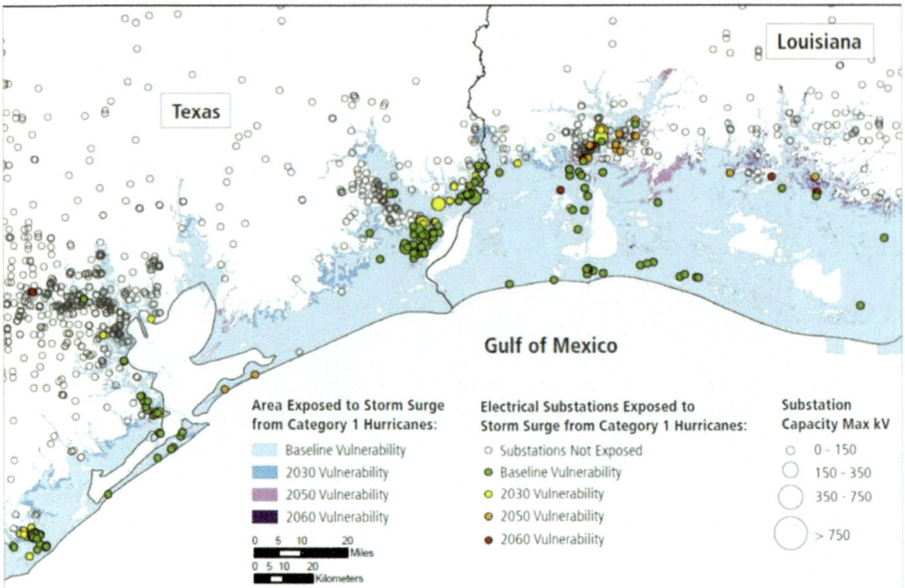

Figure 1.3 "Electricity Substation Facilities on the Gulf Coast Exposed to Storm Surge from Category 1 Hurricanes under Four Increments of Future Sea Level Rise." Oak Ridge National Laboratory, 2015.[49]

Power plants are highly vulnerable due to their concentration along coastlines. Those plants closest to sea level face threats from floods during storm surges. Hurricane Sandy completely shut down several coastal power plants that were inundated by the storm. Further inland, power plants are often located adjacent to waterways for easy access to the water they need for production. As a result, they are also often located in low-lying flood plains and are susceptible to flooding.[48]

The same is true of many other types of energy infrastructure, including substations, refineries, and fuel storage facilities. Figure 1.2 shows the evolving vulnerability of electrical substations over the next 40 years, given projections of sea level rise. Many of these facilities along the Gulf Coast are already vulnerable to even Category 1 hurricanes (indicated in green).

Conclusion

The intent of this chapter was to provide some concrete detail on the threats currently facing our energy system and, therefore, to illustrate the imperative around improving energy resilience system-wide. For the purposes of this book, it is important to remember that the realization of any of these threats is the same – the power goes out, the functioning of society slows or stops, and sometimes people die. As communities, utilities, and as a nation, we need coordinated preparation to ameliorate these threats to the extent possible, and to improve our ability to respond to them quickly when they surface. The rest of this book is designed to arm the reader with strategies to do just that.

The through line of this chapter is climate change, which has created, exacerbated, or complicated all of threats to our energy system. As we work to address these threats and build a more resilient energy system, it is important to remember that any action designed to fight climate change itself will mitigate its eventual effects and will, therefore, improve long-term energy resilience. Any project (such as a solar plus storage project) that creates an immediate energy resilience benefit and at the same time helps to fight climate change should be given extra weight in the planning process.

2

Energy Resilience Concepts

The Evolving Meaning of Energy Resilience

As the imperatives driving energy resilience efforts have evolved, so too have our conceptions of what energy resilience means. To understand the situation we are in now, it is useful to examine how these changing drivers have led to different energy resilience approaches over the past half century.

On the morning of November 10, 1965, the New York Times published an abridged 10-page version of the paper, the cover emblazoned with a rare and striking site: the New York City skyline completely dark, silhouetted against a moonlit sky. The blackout that the previous night had left over 30 million people without power across the northeastern US and Ontario. The Times had been forced to borrow the presses of the nearby Newark Evening News to get their light-weight edition out in the morning.[1] The blackout was found to have been caused by a mistake in the programming of a protective relay governing a transmission line from a hydroelectric plant near Niagara Falls. That such a small error could cause such widespread damage was a shock to energy regulators and policy makers, and it set in motion the first modern effort at reforming energy resilience practices.[2] In 1968, the utility industry formed the North American Electric Reliability Council (NERC), to facilitate coordination across the country on energy reliability. Their efforts were initially focused on the prevention of the kind of technical control issues that created the blackout in 1965.[2]

In the 1970s, the energy reliability framework changed to focus squarely on energy independence. In retaliation for American support for Israel in the Yom Kippur war, the Organization of Petroleum Exporting Countries (OPEC) drastically reduced production and embargoed exports to the United States. American oil companies like Exxon, Texaco, and Mobil were still allowed to load their tankers with oil from the cartel, but OPEC barred them from landing the ships back in the United States. These actions had their desired effect on the market, and energy prices in the United States (and much of Europe) shot up. At the time, Americans relied on oil not just to power our cars, trucks, and planes but to generate electricity. Consumers faced long lines at the gas pump and shocking bills from their utilities. Retail rates per kwh nearly doubled between 1970 and 1978.[3] The American public was understandably outraged at their own government for failing to prevent or remedy the situation. Less than two months after the embargo began, President Nixon announced "Project Independence," formally kicking off a long and ultimately futile passion among American presidents for the idea of weaning the US off of foreign oil. In 1975, the Energy Policy and Conservation Act passed, establishing the Strategic Petroleum Reserve. Jimmy Carter then made energy independence a pillar of his presidency, establishing the Department of Energy, investing in new energy technologies and installing solar panels on the White House. By the close of the 1970s, American electric utilities largely stopped investing in oil fired bulk power generation, turning instead to natural gas, coal, and nuclear power. Today, petroleum represents less than 1% of electricity generation in the United States.[4]

In the 1980s and 1990s, the deregulation of utility markets led to another change in the understanding of energy resilience. Deregulation gave independent power producers (IPPs) the opportunity to break into the vertically integrated energy market and sell energy to both utilities and consumers. The resulting proliferation of new power generation facilities brought with it challenges for utilities as they raced to install transmission lines and protect the grid in new locations. This process would accelerate in the late 1990s and 2000s, as new and intermittent generation sources such as solar and wind were added to the grid. In August 2003, a high-voltage power line in northern Ohio sagged in the summer heat and came into contact with a tree branch causing an arc and fault (a scenario that would wreak increasingly frequent and severe havoc in the decades to follow). The utility's power monitoring system failed to generate an alarm, and while they investigated the problem, several more lines sagged and faulted, putting increased load on the active lines that remained. The cascade of failures that followed caused over 50 million people to loser power, many for two days, in what was the largest blackout in North American history.[5] Congress was moved by the incident to pass the Energy Policy Act of 2005, which replaced the voluntary reliability that NERC had set out earlier with compulsory standards enforced by the Federal Energy Regulatory Commission (FERC).[2]

As we described in Chapter 1, the past two decades have brought the threat of climate change into clear focus and changed the energy resilience paradigm once again. President Obama often praised domestic oil and gas development as a pathway to energy independence while simultaneously arguing for investment in more resilient infrastructure. President Biden, however, had largely eschewed statements about energy independence before the Russian invasion of Ukraine, instead speaking explicitly about reduced emissions and resilience as more worthy goal. In remarks given at the signing of executive actions designed to address climate change, President Biden said "a key plank of our Build Back Better Recovery Plan is building a modern, resilient climate infrastructure and a clean energy future."[6]

The years leading up to the Biden presidency repeatedly demonstrated that independence from foreign fuels is of no use when our energy system is knocked out by hurricanes, droughts, floods, and fires on our own soil. Instead of seeking to achieve energy independence (as measured by energy production and net consumption), it is more practical to orient our energy strategy around resilience to the many threats that now loom. We define energy resilience as a spectrum that describes the relative ability to withstand and adapt to shocks to the energy system.

Resilience at all Levels

Unlike energy independence, energy resilience planning cannot be pursued solely as a national project with a single outcome for all Americans. Instead, energy resilience planning must be undertaken by many parties in concert, including policy makers at every level of government, utilities grid operators, and local communities and consumers. Each of these actors has a distinct role to play and unique imperatives to respond to. Actions taken solely at the local, grid, or government level are of little use without coordinated actions at the other levels. Only when each party engages in active measures to improve resilience within their own sphere of influence will we create the layers of resilience we need to be ready for the threats we face.

Local Energy Resilience

In 1977, Amory Lovins (who would go on to found the Rocky Mountain Institute) published a prescient editorial in the New York Times titled "Resilience in Energy Strategy." Even without considering the nascent science of climate change, Lovins observed that the best way to build a resilient energy system was to think local.

He wrote, "Our electrical systems are so brittle because they depend on many large and precise machines rotating in an exact synchrony and the power they generate is delivered through a frail web of aerial arteries which accident, human failure or human malice can sever... If we cannot afford failures on such a large scale, we should reduce the scale. Our centralized electrical supply cannot discriminate well between users: Electricity for a water heater, which may be unaffected by a few hours' interruption, must bear the cost of the extreme reliability required for a subway or elevator. And the grid is all-or-nothing: It must be so reliable because its failure is so catastrophic. If your oil furnace breaks down, you can go next door. But if the grid fails, there is no next door."[7]

He makes several points here that have remained salient in energy resilience discussions today. First, since so much of our energy infrastructure is interdependent, the vulnerability of the system as a whole is multiplied along with its scale. As we have seen too often, even a single tree branch brushing against a sagging power line can have widespread consequences. Second, large transmission and distribution grids are nearly incapable of prioritizing loads when generation becomes scarce. We saw this in Texas in the winter of 2021: as power plants went offline, there was little a grid operator or utility could do to direct remaining power to a hospital or water treatment plant.

Local resilience is so important because it can address these two vulnerabilities simultaneously. A local municipality can build a microgrid around its water treatment plant, fire station, or police station and potentially keep them running indefinitely when the utility feed fails. The local institution is empowered to prioritize critical loads and can shield the facilities they power from the complexity, interconnectedness, and vulnerability of the larger grid.

Even an individual institution can reasonably protect itself from shocks to the larger energy system by taking advantage of "Energy-as-a-Service" arrangements to build a microgrid or other onsite backup systems that might otherwise be prohibitively expensive. Under these arrangements, an institution such as a hospital, college campus, or even the local municipality itself can contract with a developer to build the new system with no capital cost, in exchange for an agreement to purchase the energy that the new system generates at an agreed-upon rate. This rate will increase over time but is usually designed to be competitive with the rate that the institution would have otherwise been paying the utility.

As we will discuss more in Chapter 6, there is much that a local institution can do short of building a microgrid to advance its energy resilience, starting with setting priorities and making a plan. The relative ease of decision making and long-term planning that many local institutions enjoy when compared to utilities or government make it even more important that they do their part in advancing energy resilience.

Grid Resilience

As much as local institutions can do to protect themselves during grid outages, it is still incumbent upon utilities and IPPs to make these outages less frequent and shorter in duration. Reliability has long been a metric upon which utilities are evaluated, both by regulators and customers. Reliability and resilience are related concepts but not synonymous. Reliability is typically defined as the ability of the energy system to deliver electricity where it is needed, throughout the 8760 hours of the year. Resilience is the ability of that system to address risks, mitigate their impacts, and recover from them quickly. As we discussed in Chapter 1, these risks are multiplying and evolving quickly, challenging the ability of energy companies to keep up.

The effects of climate change are turning the age-old reliability challenge of delivering energy where it is needed into a resilience challenge. Instead of simply building more peaker plants to address variability in demand, utilities now need to plan for how they will deliver electricity to customers when a gas plant is brought offline by a cold snap in Texas, or how to rethink transmission in California to minimize Public Safety Power Shutoffs and the wildfires they are intended to prevent.

As we will discuss throughout the book, there are many ways in which the actions we take to address climate change improve energy resilience at the same time. However, many of the projects that we are undertaking to address climate change present new challenges for the grid. For example, the addition of new energy generation resources – often intermittent, distributed, and relatively small – can make it more difficult for the utility to predict demand, manage distribution, and address new interconnection requests. There is even fear about a potential "utility death spiral" brought on by net metering tariffs that allow individual consumers to theoretically zero-out their utility bill by generating all the electricity they need with rooftop solar. In a contradictory trend, the addition of EVs and electrification of building loads promise to send demand for electricity to new heights, potentially challenging utilities to keep up.

Compliance with regulations, market forces, and their own sustainability goals compel utilities to add ever more clean energy resources to the grid, but the nation's energy transmission and distribution infrastructure is not yet up to the task. On a large scale, there are tremendous wind resources in many areas, such as the Dakotas, that cannot be exploited because transmission lines are insufficient to bring that power to load centers. On a smaller scale, many utilities are starting to max out on the amount of distributed energy projects they can add to their distribution laterals – which were designed for unidirectional power flow along hub-and-spoke style lines. Both scales of this problem are currently found in Vermont, where regulators have halted the addition of new solar or wind resources to the grid, even though more clean power is needed to meet the state's goals. The only reason for the moratorium is that transmission and distribution grid infrastructure simply cannot accommodate more distributed energy. Vermont is not alone, and across the country, utilities and regulators have had to curtail production from existing clean sources due to transmission or distribution congestion.[8]

Utilities need to address all these challenges if they are going to succeed in building a more resilient energy system. However, they will need to do so in cooperation with local customers, regulators, and even legislatures. Utilities have their work cut out for them, managing vegetation near lines, upgrading transformers and other infrastructure to handle new resources and growing loads, undergrounding lines, and building utility side microgrids or energy storage systems. Their customers can actively participate in demand response or time-of-use rate programs, but perhaps the best thing customers can do is to share the load by undertaking their own deliberate energy resilience planning program. Likewise, the government can provide funding for transmission and distribution grid upgrades and incentives for hardening infrastructure, but they can also help by ensuring continuity of essential services even when the grid has failed.

Government Resilience

Prolonged or widespread power outages can threaten nearly every aspect of our societal functioning, from education to economic activity and (perhaps most of all) public safety. The government, implicitly or explicitly, bears responsibility for protecting all of these. It is up to state governments and local governments to ensure that they have done the planning necessary to anticipate, withstand, and recover from failures of the energy system, whether they are caused by extreme weather events or just failure of energy infrastructure that is under increasing strain. The government is the only actor that is both responsible for and capable of the coordination necessary to ensure that their constituents have uninterrupted access to emergency services, healthcare, water, heat, and food, no matter what may be causing havoc in the energy system.

Hurricane Sandy, in 2012, set off a wave of state level energy resilience planning, aimed at mitigating the harm of such events to their constituents.

The variety of these programs speaks to the breadth of levers state and local governments have at their disposal to advance energy resilience.

In New York, the state government focused on distributed energy resources (DERs) and microgrids as the key to improving resilience of critical facilities. The New York State Energy Research and Development Authority (NYSERDA) sponsored the New York Prize program to encourage the development of microgrids through public/private partnerships and engaged in the ambitious "Reforming the Energy Vision" (NY REV) proceedings to reform utility business models to better support the proliferation of DERs.[9] Neighboring Connecticut started its own "Microgrid Grant and Loan Pilot Program" around the same time, committing $48 million to incentivize the development of microgrids to improve resilience. The New Jersey Board of Public Utilities formed the nation's first Energy Resilience Bank in 2014 with $200 million of Community Development Block Grants designated for a variety of projects including resilient CHP and microgrids.[10] In California, then Governor Schwarzenegger signed an executive order to establish cross-agency collaboration on resilience planning. Legislation followed soon after to create the California Integrated Climate Adaptation and Resiliency Program. The program codifies the resilience planning framework through which the California government will both seek to fight climate change and adapt to its significant effects on the state.[11]

We will discuss this in much greater detail in Chapter 5, but utility regulators in each state also have an important role to play. The utility industry is one of the most heavily regulated in the United States, and the public utility commissions and public service commissions in each state have a great deal of power to engender greater resilience in the energy system. Regulators have a variety of useful tools in their belt, including allowing utilities to own and rate-base resilience assets such as microgrids and performance-based ratemaking models that allow utilities to monetize improvements in resilience.

There is also a role for the federal government. FERC closed a docket in 2020 that was focused on energy resilience because commissioners felt that the topic is best tackled at a regional level.[12] However, there are resilience opportunities that are national in scope, perhaps most prominently the addition of interconnections among the country's three regional grids. The Western Interconnection, the Eastern Interconnection, and the Electric Reliability Council of Texas operate independently with few connections between East and West and very few connections to Texas. The isolation of Texas' grid compounded the effects of the 2021 freeze, and an interconnected grid would allow operators better flexibility for getting electricity where it is needed during an emergency and for the more efficient operation of markets. A National Renewable Energy Laboratory study published in 2021 demonstrated that increased interconnection in the "seams" between grids would dramatically improve our ability to share energy between regions and to send energy from large-scale renewable projects to markets where it is needed the most.[13]

More than any specific action, the most important thing a state, local government, or regulator can do is to engage in comprehensive resilience planning and to ensure that they are coordinating with power providers, utilities, emergency services, and off-takers to prepare for and respond to energy emergencies.

Finding a place on the Energy Resilience Spectrum

Any government, utility, or customer's position along the energy resilience spectrum is designated by their relative ability to withstand and bounce back from a shock to the energy system. Each of these parties needs to assess this question from the perspective of their constituents. A utility, for example, could evaluate what percentage of their customers they are able to keep powered when a storm strikes, and how quickly they can get those who lose power back online. A government could ask themselves what percentage

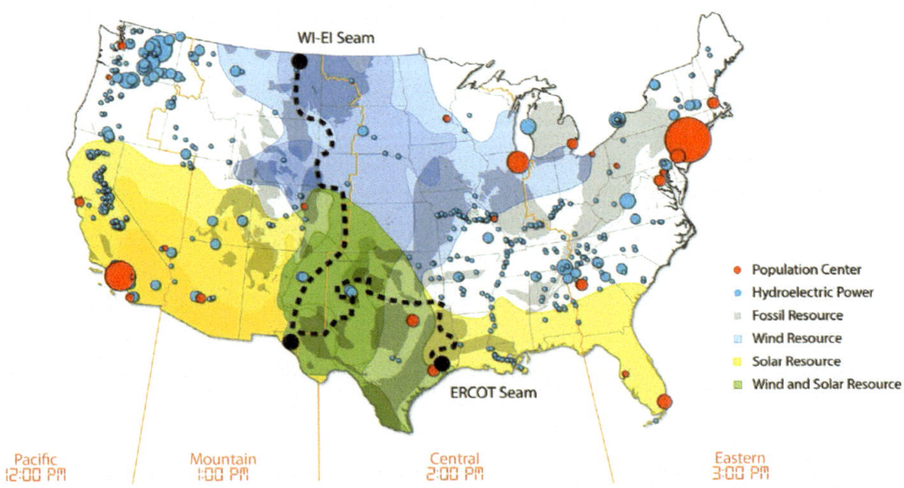

Figure 2.1 A. Bloom et al., "The Value of Increased HVDC Capacity Between Eastern and Western U.S. Grids: The Interconnections Seam Study." IEEE Transactions on Power Systems. vol. 37, no. 3. May 2022.

of the population that lives in their territory will have their needs met and be adequately cooled, warmed, and fed during such an emergency. A local institution like a military base or university can measure resilience in how well their planning protects their vital operations when the grid and regulators have not.

Thorough energy resilience planning is essential for all these actors, and their place along the energy spectrum will be determined by the execution of these plans, including the concrete measures and investments they choose to undertake. Like any investment, energy resilience measures involve trade-offs, and more is not necessarily better. Instead, each actor has to find their optimal place along the energy resilience spectrum based on the marginal costs and benefits of each potential project.

Figure 2.2 presents the perspective of a single off-taker institution, like a hospital, college campus, or local government, and displays the relative costs associated with a variety of scenarios. This energy resilience spectrum illustration is not based on a quantitative analysis of system costs but general rules of thumb identified in the authors' research.

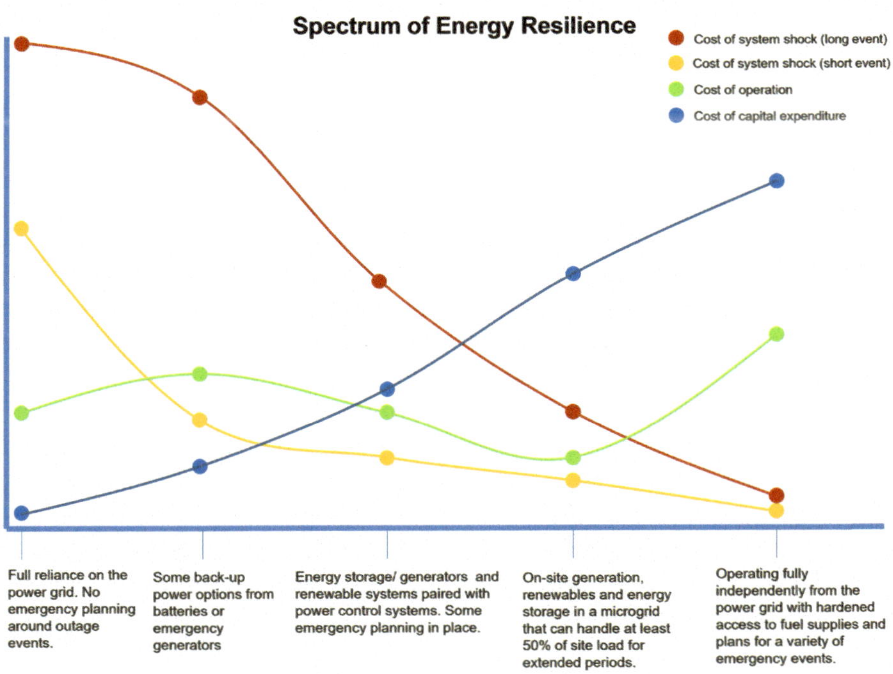

Figure 2.2 The spectrum of energy resilience for local communities or institutions.

As Figure 2.2 illustrates, the cost of system shocks (be they outage events, supply challenges, or price spikes) can far outweigh the regular cost of energy. Taking even the most basic of energy resilience measures can dramatically reduce the costs of shorter system events.

At a certain point institutions may find that pushing ever further along the energy resilience spectrum may involve complex facility upgrades or generation equipment that is expensive to install and operate. In these cases, institutional administrators must weigh additional costs against the cost of potential threats to system reliability. For some institutions, going to the right-most extreme of the resilience spectrum will make sense. For others, the optimal outcome may be in the middle or even closer to the left.

Given our definition of energy resilience as the relative ability to protect the functioning of vital operations during interruptions to the regional power supply, it is important for each actor, whether utility, government, or off-taker, to prioritize its functions and operations in terms of how much energy each demands, and how critical each is to human safety and to the mission of the institution or community. This prioritization can then be weighed against the capital costs and payback of the energy resilience projects that would address those needs. For instance, a university typically includes student housing (where electricity, lighting, and heat are vital), administrative offices, classroom buildings, research facilities, and athletic centers. If the university invests in on-site generation and a microgrid, but not one of them can provide 100% of the campus's typical load, then it will have to make decisions about which facilities receive power in emergency situations. This may include lighting and heating the dorms and temperature control at research facilities. It may only include the athletic center which can then be used as a refuge. Energy resilience planning is all about finding that sweet spot where an institution's ability to protect its vital functions from power outages converges with sound financial planning.

Of course, projects that improve energy resilience often improve financial security as well. Two of the key components of most energy resilience plans – on-site generation and operational efficiency – are often associated with significant opportunities for energy cost savings. By purchasing cheaper energy from on-site generation, and less energy overall thanks to energy conservation measures, institutions can significantly improve financial security while ensuring energy resilience and helping to protect their overall mission. The growing risks facing the energy grid described in Chapter 1 offer financial justification for a move to the right along the energy resilience spectrum for utilities, governments, and local off-takers, but just how far must be determined on a case-by-case basis through an evaluation of constituents' needs.

Just like energy resilience, environmental sustainability can be thought of as a spectrum. Sustainable design and operation are not an all-or-nothing game, and every decrease in pollution and increase in efficiency can improve environmental outcomes. However, it is important to note that although energy resilience is often associated with greater environmental sustainability, either can be achieved without necessarily affecting the other. A number of sustainability measures can be taken which have no effect on energy resilience, and vice versa. In an op-ed for The New York Times, the urban planner and developer Jonathon Rose described the dichotomy this way: "After 9/11, Lower Manhattan contained the largest collection of LEED-certified, green buildings in the world, but that was answering only part of problem. The buildings were designed to generate lower environmental impacts, but not to respond to the impacts of the environment."[14] Unless the design of a facility included a plan for protecting the flow of energy during a system shock like Sandy, an otherwise sustainable building could find itself in a vulnerable position on the energy resilience spectrum.

Without a well-known benchmarking system like LEED, and third-party guidance from organizations like the US Green Building Council that developed the LEED standards, it can be difficult for administrators and facility managers to plot a course toward energy resilience and to find a place on the energy resilience spectrum right for their community or institution. The remaining chapters of this book are designed to guide you through this process and to familiarize you with the technologies and techniques involved. For each of the three actor types (grid, government, and local institution), we define easy-to-use benchmarks to allow managers to assess their current performance and plot a practical course to greater energy resilience. However, before this planning process can begin, it is important to understand that the right position for each community will be determined by its unique operational and financial circumstances and concerns.

Conclusion

Each of the three main actors discussed in this book, local communities and institutions, grid operators, and government agencies, have their own set of imperatives and opportunities to respond to in regard to energy resilience planning. Working together, they can create the layers of resilience we need to ensure that the human impacts of energy emergencies are mitigated.

There are dollars and cents ways to identify the "right" place on the energy resilience spectrum for any institution. A utility like Pacific Gas and Electric in California must weigh the tremendous financial liability that can follow from outages like those in 2019 against the significant capital costs of vegetation management, undergrounding, microgrids, and other measures that will be required to prevent them from repeating. Legislatures need to carefully weigh each tax dollar spent and ensure that the return in public safety and economic activity justifies the expenditure. Regulators must evaluate reliability and resilience upgrades while considering their effect on energy rates.

There are, however, impacts from energy interruptions and insecurity that cannot be concretely valuated in financial terms. Success for all three of our actors should be measured in human well-being and public safety most of all. The best way to ensure that the externalities of energy insecurity are reflected in energy resilience planning is to bring all affected parties to the table. In the next chapter, we will discuss best practices for organizing constituent input and turning it into an actionable resilience plan.

Part 2

Energy Resilience Strategies and Tactics

3

The Energy Resilience Planning Process

This chapter will explore the energy planning process, from developing a business case for energy resilience efforts to the evaluation and modification of a successful plan. This process will look different for utilities, government bodies, regulators, and local institutions. In all cases, the process is at its heart one of building consensus and buy-in among stakeholders, ensuring that each project is part of a long-term institutional commitment to building energy resilience.

In the wake of Hurricane Sandy in 2012, it became clear to federal agencies that they needed better guidance on resilience in a world increasingly beset by climate disasters. In our previous book "Energy Resilient Buildings and Communities: A Practical Guide," we focused solely on energy resilience planning for local institutions and communities. Since that book was published, federal agencies have released several guides for energy resilience planning intended for local and state governments, electric utilities, and individual buildings or campuses. We summarize a few of these in Table 3.1 and will draw on them later in this chapter. It is important to note that although the specific steps may vary depending on the scope and intended audience, the various guides have much in common as indicated by the color coding and in the key for the table. We have drawn from these similarities six key steps: engaging stakeholders and setting scope, establishing a baseline, performing a risk assessment, identifying new resilience measures, planning implementation, and evaluating performance.

We will describe each of these areas in greater detail later in the chapter, including how the planning process diverges for grid, government, and local actors. Finally, we will highlight a planning resource for each of these actors.

Table 3.1 Commonalities in recent federal government resilience guides

	Guide to Community Energy Strategic Planning	Technical Resilience Navigator	Climate Change and the Electricity Sector	DHS Resilience Framework
Publisher	US Dept. of Energy, Office of Energy Efficiency & Renewable Energy	Federal Energy Management program, NREL, PNNL	US Dept. of Energy, Office of Energy Policy and Systems Analysis	US Dept. of Homeland Security
Audience	State and Local Governments	Site Level End Users	Electric Utilities	Federal Government
Step 1	Form Leadership Team	Site Level Planning	Scope the Resilience Plan	Engage Stakeholders
Step 2	Identify/Engage Stakeholders	Baseline Development	Develop Inputs for Vulnerability Assessment	Identify Critical Mission
Step 3	Develop Energy Vision	Risk Assessment	Determine Exposure of Assets and Operations	Conduct Criticality Assessment
Step 4	Assess Current Energy Profile	Solution Development	Estimate the Consequences of Impacts	Assess Liabilities
Step 5	Develop Goals and Strategies	Solution Prioritization	Assess Vulnerabilities	Identify Resilience Gaps and Solutions
Step 6	Identify/Prioritize Actions	Roadmap to Action	Identify and Assess Resilience Measures	Integrate Resilience Readiness Solutions
Step 7	Financing Strategy		Build Portfolio of Resilience Measures	
Step 8	Develop Blueprint for Implementation		Monitor Evaluate, and Reassess	
Step 9	Plan to Evaluate			
Step 10	Adopt and Publicize			

Key
- Engage stakeholders, establish scope
- Assess energy resilience baseline
- Perform risk assessment
- Set goals and idnetify resilinece tactics
- Plan implementation of new tactics
- Execute measures and evaluate performance

Chapters 1 and 2 of this book describe the risks faced by our energy system and how our understanding of resilience has evolved to address them. Before delving into the detail of each phase of the energy resilience planning process, energy managers will need to build a business case to justify the planning effort. As with any considerable investment, energy projects require a strong business case – not just to get authorization for funding but to ensure that all parties involved understand the value of the project and are invested in its success. The foundation of this business case can show the quantified risks facing an institution's energy supply or infrastructure. Risk assessment will be covered in greater detail later in the chapter. It is up to the energy resilience "champion" in each organization to describe these risks to their stakeholders in a way that will resonate and justify convening an energy resilience planning process in the first place.

It is not necessary to identify specific solutions ahead of the formal planning process. In fact, allowing other stakeholders to play a role in identifying the solutions will go a long way toward developing buy-in and long-term support. The task at this point is to identify the threats, quantify their impact on the institution, and illustrate a broad vision in which the community can solve this issue and become better prepared for whatever challenges lay ahead.

There is not a single path to success, and your approach should be tailored to the needs, interests, and strengths of your institution. Below are some hypothetical examples of how communities might approach this differently.

Rallying Around Public Safety

Kate is an urban planner for a town in New York. After a major storm event, there have been several articles in the local paper and comments at town hall meetings about the fragility of the town's energy infrastructure. City officials seem responsive to these concerns and want to ensure a safer future. With this foundation, Kate can build a business case around the risk posed to the community by future weather events. She can demonstrate that by investing in a resilience project, the town can protect its citizens, avoid the costs of lost productivity, and demonstrate forward-thinking leadership in response to community concerns. Kate reaches out to the Mayor, the city council, and the Chamber of Commerce. She emphasizes human impacts in this proposal, showing how efforts in energy resilience can make the community safer for residents. She also stresses that this investment will make the community more attractive for businesses and individuals considering a move to the area. Understanding that local budgets are tight, Kate explores opportunities such as energy-as-a-service, energy performance contracts, and other innovative financing approaches so that when someone inevitably says, "We can't afford it," Kate can show them ways it can be done with minimal local investment.

Stressing the Green Benefits: Environmental and Financial

Andrew is the head of operations for a college in California. The college has made public pledges to reduce energy usage and improve the sustainability of its operations. The student body also wants a greener campus and has a vocal environmental club. Andrew develops a business case focused on the positive environmental impact that on-site generation, energy efficiency, and even emergency planning can have on the campus and how that could help the college meet its environmental goals. He couples this with analysis of potential cost savings to help justify capital investment. Andrew speaks with the school president, the provost, some environmental faculty members, student organizations (the environmental club and the student government), and possibly even representatives from local businesses or on-site vendors. Andrew can work with the school's finance department to find a way to borrow from future energy budgets to finance projects today.

Using a Mandate to Leverage Better Resilience Planning

Jonathan is the energy manager for a military base in New Mexico, and, in this role, he must follow executive orders from the White House to hit certain energy reduction targets. His base commander has also expressed concern about combat readiness should connection to the local utility fail. Jonathan knows that his efforts cannot be seen as a distraction from the core mission but is aware that he can leverage executive orders and a risk-averse culture to make changes. As he discusses the business case, he will focus on what might happen to the base and its combat readiness should they experience widespread power loss or noncompliance with an executive order. Jonathan engages commanders on base, the local community's planning or sustainability office, and businesses that operate on base. He can work with the Office of the Assistant Secretary of Defense for Operational Energy to obtain financing to support these kinds of measures.

In each of these cases, the champion of energy resilience had to understand the drivers for their institution, the metrics by which that institution will measure success, and the stakeholders that must buy into this concept before progress could be made on a formal resilience planning effort. Once this effort is officially convened, it should pursue the following six key steps.

Step 1: Engage Stakeholders and Establish Scope

Once you have completed your analysis and can effectively convey the business case of your energy resilience projects, the next steps are to present your case to stakeholders to build a coalition of support and convene a planning session.

The stakeholders should include the groups and individuals who:

1. have decision-making power over energy or capital planning at your organization, or influence sufficient to help or hinder the process;
2. have expertise that will be valuable in the planning process; or
3. represent a key constituency of your organization, such as customers for a utility, students at a university, or local business and community groups in a municipality.

In some cases, you might want to build support for the energy resilience idea in general, using meetings with stakeholders and your customized business case information to discuss the importance of energy resilience. In other organizations, the best approach might be more direct: getting the right players in the same room to determine the importance of energy resilience and outline parameters for developing an energy resilience plan.

The Department of Defense's Net-Zero Military Installations planning guide[48] offers a valuable list of participants for this kind of effort.

- Installation management (such as the base commander)
- Energy manager
- Facilities and maintenance personnel
- Fleet vehicle manager
- Director of public works
- Contracting officer
- Environmental manager
- Master planning
- Installation public affairs officer
- Installation security officer
- Utility company

For a utility, the list of stakeholders will likely include customers, community groups, subject matter experts, and government officials. The regulator may also convene stakeholders to inform the commission's guidance to utilities on how to develop new resilience programs. Stakeholder input will inform a report on energy resilience from the commission, which is then delivered to the utility. When the utility then submits a rate case related to energy resilience, they will have incentive to make sure it reflects stakeholder feedback in hopes that it will, therefore, be received favorably by regulators.

As you organize stakeholders, you should leverage the business case pre-planning you did. For example, when speaking with local businesses, you could talk about the impact other natural disasters have had on local businesses and demonstrate the value of this engagement to protecting their interests. Some research on actions other communities have taken in this area can be valuable in demonstrating that these concepts are already being applied.

Rather than simply recruiting as many participants as possible, direct your efforts to those parties who are most likely to move the process forward and make the final project a success. An energy resilience planning effort can suffer from too much involvement as easily as too little. In the case of a municipal government, you will most likely want to involve representatives from the community. This can be done at regularly scheduled town hall or committee meetings. It may be useful to establish a baseline at these meetings with a brief presentation of the threats to the energy system, the economics of energy disruptions, and some discussion of solutions including energy efficiency, energy storage, and on-site generation.

If you do not have a mandate from senior leadership to move forward with an energy resilience planning effort, you may need to convene stakeholders to make the business case and garner feedback on how to proceed. The first goal of this meeting is to understand the opinions of each stakeholder on the importance of energy resilience and what they would be willing to contribute to efforts on that front. The second goal is to get all parties to agree on a path forward.

Ensure a leadership team is named early in the stakeholder engagement process. This small group will be responsible for moving the process forward and executing the plans developed by stakeholders. This group should include the energy director or whoever leads energy work in the organization and other parties that can share the workload and to engage different groups as the plan develops. The members of the leadership team, serving as the project owners, will need to meet regularly to make sure this project maintains momentum.

Attempt to represent each of the major groups that will be involved with implementation of the final plan. Unlike with your stakeholder group, this team should include only people with decision-making authority within the organization.

Step 2: Assess Energy Resilience Baseline

Once you convene the stakeholders and identify a leadership team, you should work to establish a performance baseline as one of your first activities as a group. There are two main elements to this process: 1) assess the criticality of existing infrastructure and the operations they support, and 2) identify any energy resilience measures that are already in place.

The process of conducting a "criticality assessment" is described well in the DHS Resilience Framework.[1] The authors of that guide describe the evaluation of operational or business value for a particular asset to be a function of both the essential functions that rely on that asset, and how sensitive those operations are to its loss.[1] Electronic life support systems in a hospital, for instance, are highly critical for obvious reasons and rely on the electrical distribution system of the hospital. Life support systems can tolerate very little downtime and are, therefore, very sensitive to potential utility blackouts. The DHS guide provides the following four definitions for "continuity criticality." These levels were defined within the context of the DHS mission but should apply at a high level for any government body, and the concept is useful for any institution.

The next step in the baselining process is to evaluate how resilient the asset is to disruption. Returning to our hospital example, there may be a diesel backup generator for the most critical loads, providing some measure of resilience to a utility blackout. Perhaps, certain pieces of equipment carry battery backup on board. Cataloging these measures and their capability to keep critical functions running is essential to the risk analysis performed in the next step. The assessed risk will be equal to the difference between the threats facing critical functions and the existing ability to withstand and recover from them.

The adage "you can't manage what you don't measure" is so fundamental to sound energy management that it has become something of a cliché. An energy resilience baseline effort determines the current state of your technology, infrastructure, skill sets, and operating procedures in each area of energy resilience. The next chapter in this book provides a maturity model template for your organization to use in understanding the current state of your energy resilience approach.

Table 3.2 Continuity criticality quantitative scoring definitions (Table D-2 from *DHS Resilience Framework*)[1]

Continuity Criticality Level 4	**Very high consequence:** Loss or disruption of the asset or function has exceptionally grave consequences; such as extensive loss of life, widespread severe injuries, and total loss of primary services, core functions, and processes.
Continuity Criticality Level 3	**High consequence:** Loss or disruption of the asset or function has grave consequences; such as loss of life, severe injuries, loss of primary services, and major loss of core processes and functions for an extended period.
Continuity Criticality Level 2	**Medium consequence:** Loss or disruption of the asset or function has moderate to serious consequences; such as injuries or impairment of core functions and processes.
Continuity Criticality Level 1	**Low consequence:** Loss or disruption of the asset or function has minor consequences or impact; such as a slight impact on the core functions and processes for a short period of time.

Step 3: Perform Risk Assessment

Once you establish your baseline, you can identify the gaps between your current energy resilience and the level of resilience you need to achieve to protect your critical functions. Start with the criticality assessment performed in Step 2. For each critical function that your organization performs, trace it back to the infrastructure on which it relies and then examine the vulnerabilities of that infrastructure to energy interruption.

The hospital discussed above may trace its life support systems back to a particular electrical circuit, which is, in turn, protected by a backup generator. However, if this generator is in the basement, it may be vulnerable to flooding, even if it was determined that floods were not a significant risk when the generator was installed. As we discussed in Chapter 1, climate change is changing the risk profile for nearly every piece of infrastructure. Even if a hospital was built outside of a 100-year flood plain, it may now need to make flooding a part of its resilience planning process. The Department of Energy (DOE) Guide for Climate Change Resilience Planning for the electricity sector makes this explicit and describes a detailed process for how to identify climate change impacts on utility infrastructure, and how to include that data in a risk assessment.[2] The DOE guide describes reputable sources for climate impact projections, including heat, rainfall, flood risk, and others. It also includes tools that can be used to assess risk to physical infrastructure based on these projections.

In their description of risk assessments, the Department of Homeland Security includes an evaluation of "what can happen (hazards and outcomes), the likelihood of it happening (the combined probability of hazards and vulnerabilities), and the consequences if it does happen (severity of outcomes)."[1] The example above pertains to natural hazards, but you should be careful to include possible technological and human hazards as well, as described in Chapter 1.

For each of these threats, evaluate the likelihood that they will impact your organization based on publicly available tools and resources and consultation with experts. Finally, determine the severity of this potential impact on your operations as a way to prioritize energy resilience goals.

Chris Castro, the Director of Sustainability & Resilience for the City of Orlando, told us about how the city's risk assessment helped to guide their efforts: "We started with a vulnerability and risk assessment. We did this using FEMA and NOAA guidelines on hazards. We looked at the probability impact, and frequency trends over time. We wanted to understand how each of these hazards could impact the city and its operations. This allowed us to focus. We worked closely with our emergency management team to understand our current plans and walk backward to think about ways we can enhance those plans and improve our resilience as a whole."[3]

Step 4: Set Goals and Identify Resilience Tactics

Once current risks are evaluated, your organization must set goals for what level of energy resilience it wants to achieve. In Chapter 2, we noted that energy resilience is a spectrum, and any organization's need to protect critical operations must be weighed against the capital and human resources required to take on additional resilience measures.

It may be useful to summarize your energy resilience goals in a vision statement, to ensure that each goal aligns with your organization's long-term planning. This vision could be set early in the process during the stakeholder engagement phase, but waiting until after the risk evaluation will allow you to inform the energy vision with real data and intelligence about the problem to be solved. The vision needs to be realistic and functional. It should be flexible enough to allow the planning team to evaluate multiple approaches. When developing language for this vision, imagine how a group of strangers would develop your actual energy resilience plan based only on the vision language. Would they know what you wanted to do and when you wanted it done?

Example vision statement	Evaluation
Corellia Township will utilize cutting-edge technologies to become a net-zero community and the most energy resilient town in the United States.	This is too ambitious. The planning team will likely soon find that achieving this vision is impossible, leaving them with no functional vision.
Barnett College will evaluate energy resilience opportunities and attempt to reduce energy consumption by 10%.	This is not ambitious enough. It focuses on process over results and is prescriptive in terms of what will be addressed. If energy efficiency is your goal, that 10% figure may make sense (though it is very modest). If energy resilience is your goal, efficiency may or may not be part of the strategy.
Kimball Medical Center will employ highly efficient cooling and lighting technology combined with a microturbine generator to allow 80% of facilities to operate in island mode when necessary.	This is too tactical. It ties the vision to specific technologies. It sets specific technical goals that are likely unknowable ahead of time.
By 2025, Fort Deckard will be capable of supplying energy to 85% of its operations even when power is not available from off-site sources.	This is a good energy vision statement. It is ambitious but realistic. It includes a timeframe. It is clearly focused on reliability.
The Devlin-MacGregor corporate campus will leverage advanced technologies to reduce energy use by 20% by 2030, demonstrating our role as a global leader in sustainability.	This is also a good energy vision. It is clearly focused on risks associated with energy costs, but it also ties this effort directly to the corporate mission.

Once an energy vision is established, you can set specific goals for energy resilience. It is important to have active stakeholder involvement in this stage via a facilitated goal-setting meeting. The output can be a final set of agreed-upon goals or a series of goal recommendations for the ultimate decision maker to address (depending on your process). It is not important to work out technical details of every tactic in the plan. Instead, focus on capturing and understanding the priorities of the stakeholders. Below are some examples of energy resilience goals. In each case, the goals are paired such that achieving one will help achieve the other.

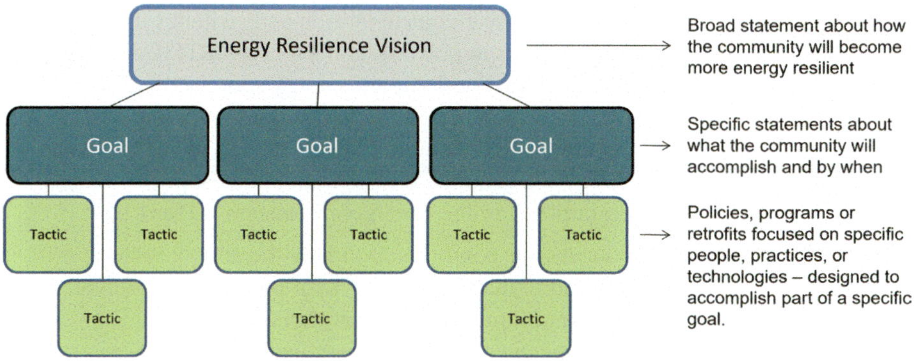

Figure 3.1 Energy resilience plan structure.

Type of goal	Examples
Energy purchasing	• Negotiate energy purchasing contracts so that 75% of the energy purchases for the next 10 years are at a fixed, predictable rate. • Reduce our cost of purchased BTU by 15% by [DATE].
Energy efficiency	• Fund identified energy conservation measures with a payback of three years or less. • Reduce building energy consumption by 20% within five years (using the previous calendar year as a baseline).
Transportation *(more applicable to municipal governments)*	• Improve deployment of bus routes by [DATE] so that no member of the community would need to walk more than one mile to reach a bus heading downtown. • Replace 50% of fleet with EVs by [DATE].
On-site generation	• Develop on-site generation that can handle 50% of our institution's electricity load for at least three weeks in the case of a grid failure by [DATE]. • Eliminate our reliance on delivered electricity within ten years.
Renewable energy	• Deploy 20 MW of renewable energy generation by [YEAR]. • Secure 30% of our energy from on-site renewable energy sources by [YEAR].

Microgrid	• Deploy a power control system and switches that will allow us to power [LIST OF 2-4 CRITICAL FACILITIES] with on-site energy sources by [DATE]. • By [DATE], deploy a microgrid that will allow our institution to operate in "Island Mode," providing what energy we generate on-site to the highest priority loads (as outlined in the Emergency Response Plan).
Emergency planning	• Educate our community on how the energy systems will respond in the case of various power outage events so that community leaders can plan accordingly. • Integrate emergency response and power control activities to support critical services within 30 minutes of a blackout event.

Once a set of goals is finalized, the leadership team should develop a set of tactics to achieve those goals. While the analysis is likely to identify the cost effectiveness of a tactic, it is important to also focus on how well each tactic is matched to your specific goals. Below are some examples of specific tactics you might consider and what metrics you could use to evaluate them.

Vision	Example goal	Tactic	Evaluation metrics
Energy independence	Develop on-site generation to handle 50% of our institution's electricity load for at least three weeks in the case of a grid failure by [DATE].	Build a 25–35 MW gas-fired cogeneration plant next to the existing central plant.	• Cost to develop the plant • Anticipated payback based on expected gas rates and electric utility rates • Percentage of the institution's load that can be handled • Security of the natural gas supply

Insulation from price spikes	Negotiate energy purchasing contracts so that 75% of our energy purchases for the next 10 years are at a fixed, predictable rate.	Secure bids from third-party energy suppliers to evaluate savings options.	• Likelihood of electricity prices rising or falling. The most reliable source for this kind of information is the U.S. Energy Information Administration • The rates that our organization is able to negotiate and how this compares to current energy rates • Potential cost savings compared to another approach (such as developing our own on-site generation)
Environmental sustainability	Reduce overall greenhouse gas footprint by 20% in the next 15 years.	Leverage energy performance contracting to reduce energy consumption in buildings 20%.	• Possibility of a 20% reduction given past efforts and current technology • Ability to use internal capital to finance energy performance improvements and thereby retain more of the savings • Planned changes to the building portfolio that might affect this effort

At this point in the process, you will have developed an overall energy resilience vision, goals to help you achieve that vision, and tactics you will employ to achieve those goals. Starting with a vision and goals ensures that all participants can evaluate tactics using the same parameters for success.

Step 5: Plan Implementation of Resilience Tactics

Once the set of tactics is matched to your resilience goals, the leadership team should create a clear and specific plan for their implementation.

For each tactic, the team needs to identify who will be responsible, how the project will be funded, and a timeframe for its execution. As always, stakeholders should be engaged so that they know what to expect, how they may be impacted during implementation, and how each tactic will benefit them. Once complete, the implementation plan should describe a timeline of actions that will get the organization from current status to its goals.

It is important that this plan include detail on how each of the tactics will be successful. This includes the clear assignment of responsibility and all other roles necessary for execution. Crucially, the plan must also be clear about how each tactic will be funded.

Although you may identify a few low-cost ideas that you can execute in an afternoon, most of the tactics you develop will require investment. There may be one funding mechanism chosen to cover all tactics, but it may be valuable to think about each tactic individually to understand the best financing approach. The wide variety of financing mechanisms available today means any institution can find some funding options to make progress on energy resilience. Here are several approaches:

Internal capital: For certain projects, the institution may be able to dedicate part of its annual budget to pay for projects.

Pros: This traditional approach ensures that the institution reaps 100% of the rewards from their efforts. This can be a good route for projects that create energy resilience but do not produce an income stream for an outside party, like energy savings via an EPC, or sale of energy via a PPA.

Cons: Securing large allocations from annual budgets is typically very difficult. Using the annual budget process may delay action.

Borrowing from energy savings: For projects that are projected to reduce energy costs, the finance department may allow for money to be borrowed from those future savings in order to pay for projects that generate them. Municipal governments can also use bond measures to employ this approach.

Pros: Creates a budget-neutral approach that keeps all savings in-house.

Cons: Should projects fail to secure anticipated savings, for example, energy budgets will fall short. This approach can only be used for tactics focused on reducing energy consumption.

Grant funding: The institution works with government agencies or non-profit organizations to secure program funding or grant funding to pursue these projects.

Pros: Can be used to pursue any kind of project, including projects that do not yield actual dollar savings.

Cons: Requires an investment of staff time to pursue grants or Federal program dollars. Also, grants often only cover part of an effort.

Leveraging private investment: The institution can work with private businesses (such as vendors, tenants, landlords, utilities, etc.) to encourage activities that support the energy resilience plan. This may simply involve convincing the businesses of the value of these activities. Municipal governments can use options such as preferential zoning. Companies, universities, and military bases can negotiate these activities into vendor contracts.

Pros: Generates action on energy resilience without requiring internal expenditures.

Cons: Programs created this way are largely out of the control of the leadership team. Performance toward goals may suffer in cases where partners deploying the tactics are simply doing it to satisfy a requirement.

Energy service performance contracts: Private engineering firms conduct energy audits of a customer's facilities and identify energy savings opportunities. The firm then performs the identified retrofits using their own money, and the client pays for this service by giving the firm a portion of the energy cost savings over a pre-determined period.

Pros: Creates an easy, low-risk avenue to achieve energy savings without waiting on internal funding.

Cons: The institution will not get the full benefit of the energy savings. This can only be used for tactics focused on reducing energy consumption. The value of these projects appears as debt on the balance sheet.

Saved power purchase agreement: In this model, a third-party firm replaces inefficient energy-using equipment with their own, more efficient equipment at no cost to the owner. This retrofit typically includes monitoring capacity that allows the firm to track actual energy saved by the new equipment and compare it to a baseline of energy usage by that equipment prior to the retrofit. The institution agrees to buy a certain number of avoided kilowatt hours (based off that pre-retrofit baseline) at a set price (typically lower than the current utility rates). If no energy is saved, no payments are made. Once all the agreed-to-be-saved kilowatt hours have been purchased, the firm abandons their technology in place.

Pros: This approach creates an easy, no-risk avenue to achieve energy savings without waiting on internal funding. The institution is simply purchasing a service so that the cost of the project does not appear as debt on the balance sheet.

Cons: The institution will not get the full benefit of the energy savings, as some profit must be made by the third-party firm. Due to the complexity of the approach, this kind of project typically only works well for organizations with a large energy spend, where the investment in contract development and performance monitoring equipment will be justified.

Revolving loan fund: In this model, the institution allocates money into a fund that is then used to pay for energy performance projects. The savings from those projects go back into the fund and can be used to support other projects in the future. The goal is to create a virtuous cycle where savings from past projects will fund the capital costs of future projects. The use of these revenue sources must be strictly dedicated to energy projects so that other priorities within the institution are not allowed to drain this fund and defeat the approach.

Pros: Revolving loan funds have proven to be one of the most successful approaches for supporting continuous energy savings programs. This requires no investment after the initial seed funding.

Cons: This model requires a sizable up-front investment. Projects must be well evaluated and monitored to ensure savings are realized in order for the fund to properly sustain itself. This can only be used to fund energy saving measures.

This graphic from Jen Weiss' article "Revolving Credit – All Grown Up"[68] shows how a revolving loan fund works:

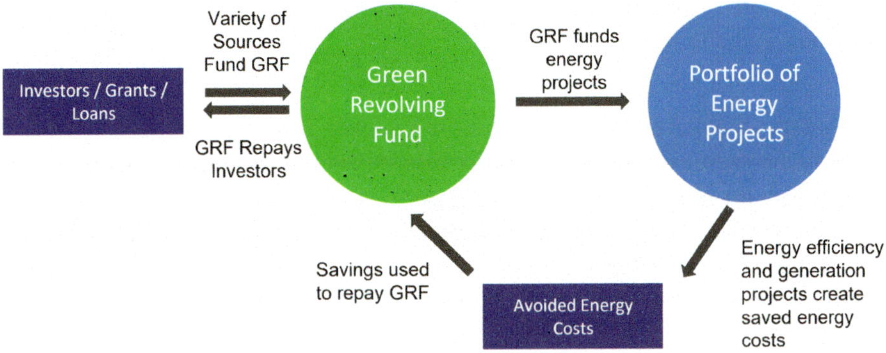

Figure 3.2 Green revolving fund.

Energy-as-a-Service: Unlike several of the instruments described above, the general category of Energy-as-a-Service, or EaaS, can be used to fund a variety of projects and not just energy savings projects. Typically, these arrangements take the form of a PPA and are centered around one or more on-site generation assets. Microgrids, for example, can be funded through an EaaS contract with a developer. In such a project, the developer pays all capital costs for the microgrid, including distributed energy resources (solar PV, battery energy storage, CHP, etc.), efficiency upgrades, and any required energy distribution infrastructure. The host institution reimburses the developer through the purchase of the energy from the generation assets included in the project. The developer may also include an energy performance contract as part of the overall EaaS agreement if energy efficiency is a significant part of the microgrid project.

Pros: The lack of capital expenditure associated with EaaS agreements puts ambitious energy projects within reach for organizations that do not have the resources on hand to fund such a project themselves. The rate structure is contractually set so that the host institution is shielded from fluctuations in energy costs and can make these costs a predictable part of annual budgets. Comparisons to existing energy costs are clear so that it should be easy for the host institution to determine if the EaaS project is cost effective.

Cons: The financial benefits of the project must be shared with the developer, who must make a margin on their investment. An EaaS arrangement will not work financially for all projects. Typically, successful EaaS projects serve a large load, with higher current energy costs and an excellent potential for new onsite energy generation.

Third-party incentives: Whatever funding source is leveraged, look at every source of financial or service support that exists. Many utilities have goals around demand reduction and provide program or financial support to their customers on this front. There are incentives and rebates available from both the Federal and state governments for a variety of efficiency and on-site generation projects.

In deciding on your financing approach, consider how that mechanism can be used to generate the maximum possible impact on the energy resilience of your institution. One best practice is bundling measures into larger projects. Projects with higher internal rates of return can help to offset the lower IRR of projects that may have other valuable benefits when it comes to energy resilience. For example, if your institution requires 20% rate of return for capital investment, you could pair a lighting upgrade with a 50% internal rate of return with a power storage project that has a 10% rate of return. Bundling projects also helps when you are dealing with projects that have IRR that is harder to define or that involves more uncertainty. Focus on creating the largest possible portfolio of activities within your available financing mechanism – mixing rates of return, ancillary benefits, and levels of uncertainty.

Step 6: Execute Measures and Evaluate Performance

Once your team develops your energy resilience plan and a financing strategy, publish the plan so that it can be easily shared with parties both within and outside of your institution. This may seem like a step that does not warrant much explanation, but there are a few considerations that will help ensure that the plan is understood and received well by your target audience.

First, be sure to write up the plan as simply and clearly as possible. Imagine that your entire team wins the lottery and departs for your newly acquired private islands. The team that will replace you needs to understand the energy vision for your institution and understand how those tactics are to be deployed in support of that vision. You should also write this plan for a non-technical audience, to the extent possible. This plan is to be your constituents' doorway into what is being done to ensure the energy resiliency of their town, their base, their campus, or their hospital.

Second, get this plan officially "signed" by the senior official in your institution and have that person share this plan with the entire community. This leader's seal of approval will provide the leadership team with leverage as they attempt to implement the plan. Also, letting the entire community know about the plan will foster goodwill toward your effort, prepare them for changes they may see as a result of it, and it may even generate some new ideas or new allies as your vision, goals, and tactics are shared with a wider audience.

Any energy resilience plan represents the best thinking of your team at that time. However, times change, and new information allows us to refine our approach. Your leadership team should be evaluating the progress of each project initiated by the plan on a set cadence. Is it progressing at the expected pace? Are the expected results being achieved? Are those results driving the goals as expected?

The easiest way to accomplish this is to simply build measurement, verification, and a formal review of your energy resilience efforts into the plan itself. The lower-level tactics should be reviewed regularly while the higher-level strategies and goals should be reviewed only after some time has passed to provide insight into how the plan performed. Specific tactics and projects should likely be reviewed at least annually to ensure that they are on track and yielding the intended results. Goals may be addressed annually or every few years. The energy vision should hopefully remain the same until it is achieved, but there may be cases when the leadership team feels the need to alter course and revisit vision.

As you have seen in this chapter, effective energy resilience planning is quite similar to other community planning projects. It involves convening critical stakeholders, developing a tiered approach toward a single vision, and leveraging your institution's strengths to make progress and build upon that momentum. Energy systems are complex and interdependent, which is why they sometimes fail and why that failure can be such a problem. The key is to attack the problem systematically and leverage advances in one area to support advances in other areas. Consider, for example, the way better lighting can lower cooling costs or the way an energy storage project can help make solar power more cost effective. If your organization can successfully engage energy systems as a whole, energy resilience efforts will yield better, more immediate results.

Resilience Planning for the Grid

Most of the information in this chapter is relevant to all three of our key actors: utilities, government, and local institutions. However, many details of how decisions are made apply more to a city government or university campus than they do to an electrical utility, which needs to engage its customers, regulators, community groups, and potentially even legislators in its planning process. Utilities and IPPs may not have access to the full gamut of funding mechanisms listed above, but they are increasingly exploring new options beyond simply rate-basing infrastructure investments. As DOE observed in their planning guide, these new mechanisms may include "cost deferral, rate adjustment mechanisms, lost revenue and purchased power adjustments, formula rates, storm reserve accounts, securitization, customer or developer funding/matching contributions, federal funding, and insurance."[2]

Utilities and IPPs should start with planning resources written specifically for them, such as the guide from the DOE mentioned earlier: "Climate Change and the Electricity Sector: Guide for Climate Change Resilience Planning."[2] The guide describes the process of assessing the vulnerability of electric utility assets to the effects of climate change, including rising temperatures, rising sea levels, and more frequent and severe extreme weather including flooding and drought. The guide is particularly thorough on the topic of climate risk to physical infrastructure and helps the reader make informed choices about how to use climate models and projections to assess this risk. The guide also provides useful suggestions on potential resilience measures, including hardening existing assets and operational changes, and how to select the right ones given the results of a vulnerability assessment.

The DOE guide also features case studies that illustrate how this planning structure has been put into action in the real world. One such study describes the effort by National Grid to assess the vulnerability of its sub-stations to flooding. The utility located its sub-stations on Flood Insurance Rate Maps and identified them as high, medium, and low risks based on flood plains. This categorization helped them to identify more aggressive measures for high-risk equipment (including plans to relocate or retire certain sub-stations) and more moderate measures for equipment with medium risk.[2]

As Figure 3.3 illustrates, the authors of the guide view the process as a cycle that repeats continuously as the landscape of hazards changes.

A detailed discussion of the particular resilience challenges facing utilities and IPPs can be found in Chapter 5, "Energy Resilience at the Grid Level."

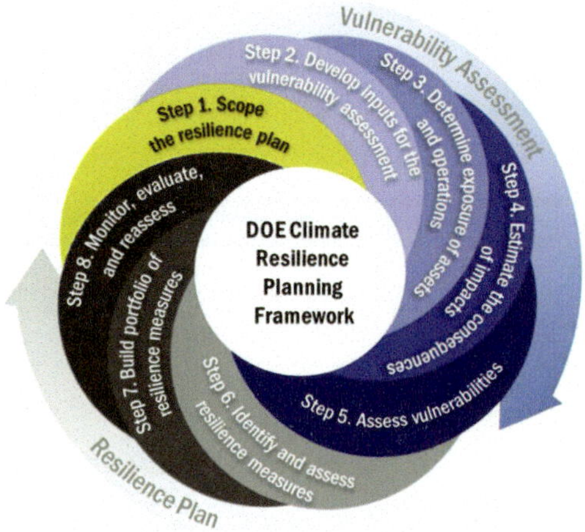

Figure 3.3 "Steps for conducting a vulnerability assessment and developing climate resilience solutions."[2]

Resilience Planning for Government

The set of stakeholders for a government resilience project is broad and includes all the constituents of the particular government body. The challenge for government is to identify the groups within this constituency that are most at risk and those with the most to contribute to a resilience planning process and convene these parties without leaving anyone out. For this reason, it is important that a government body continuously engage stakeholders throughout the energy resilience planning process.

The DOE-sponsored Community Energy Strategic Planning Academy was offered to local energy planning officials in 2011. One of authors of this book, Brian Levite, was part of the small team of energy planning experts that developed this framework and associated course materials for the Academy. The planning guide that came out of this effort, "Guide to Community Energy Strategic Planning," continues to be a valuable resource for local governments beginning the process of energy resilience planning.[4] The long-term commitment to stakeholder engagement is reflected in the planning timeline from the guide, below.

The process of energy resilience planning in a federal agency is represented well in the "DHS Resilience Framework," from the Department of Homeland Security.[1] As illustrated by Figure 3.5, the overall process is very much the same as at the local government level. However, the scope and scale of the project must include facilities and assets that are distributed across the entire nation, and, in the case of DHS, a critical mission explicitly focused on national security.

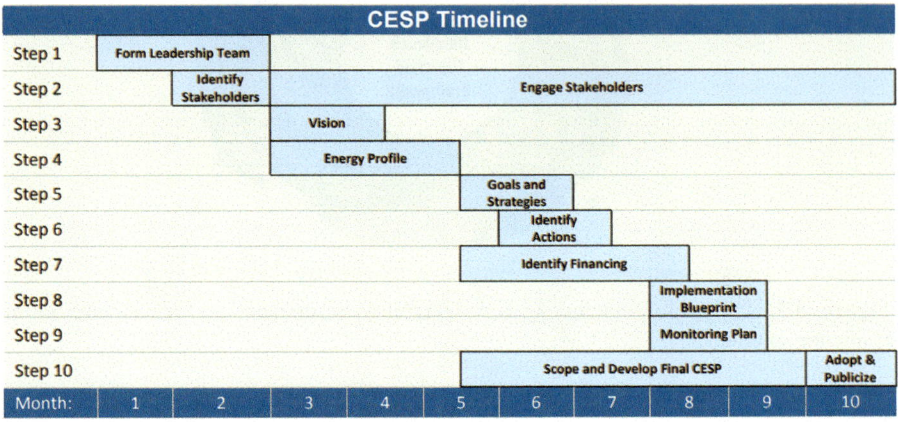

Figure 3.4 "Community Energy Strategic Planning Timeline."[2]

Figure 3.5 "Resilience Framework Process, US Dept. of Homeland Security."[2]

The case study below outlines the planning effort undertaken by Fort Bliss in El Paso, Texas. U.S. military bases have been among the most aggressive communities in pursuit of energy resilience. Although we may think of the U.S. military as an extremely hierarchical organization where priorities are identified and then executed, in truth, military groups can respond to official directives in a variety of ways. Determining the correct course of action requires discussion, expert advice, and business case development. Fort Bliss is an excellent example of how an energy resilience effort was carefully considered and deployed in partnership with numerous stakeholders.

Case Study: Fort Bliss, El Paso, Texas

[BJ Tomlinson, Energy Branch Chief, Fort Bliss, Interview with author, conducted July 1, 2013]

Energy resilience efforts at Fort Bliss serve as an excellent example of leveraging stakeholder engagement to achieve energy and water goals. The process began with directives from Washington that the military should reduce energy consumption. This call was taken up by Fort Bliss Commanding General Howard B. Bromburg, who felt conservation and efficiency were important values to champion. His successor, General Dana Pittard, championed the same priorities after Bromburg's departure. However, the enthusiastic support of these two leaders was not enough.

Staff like Energy Branch Chief B. J. Tomlinson had to engage a wide variety of stakeholders in order to achieve the goals set forth by the base commanders. Tomlinson's organizing activities included:

- *Partnering with the Pentagon's newly created net-zero energy program and volunteering Fort Bliss to be a pilot installation to receiving technical support on program development.*
- *Working closely with local governments – Fort Bliss has a footprint that spans both Texas and New Mexico. As new projects were considered (particularly new power generation projects), the local governments and utilities were at the table creating a project plan that was valuable to all involved. These interactions were not always smooth, but these relationships are improving over time and Chief Tomlinson believes that finding common ground on new projects is getting easier.*

- *Working with energy performance contractors to complete efficiency projects when additional capital funding for efficiency projects was not available.*
- *Enlisting the support of the base commander to circumvent red tape and fast-track projects with clear energy resilience and efficiency benefits.*

These collaborations have delivered real value to Fort Bliss. The base has implemented a wide variety of energy saving measures in their 2200 buildings, encompassing 32 million square feet and representing a $357 million annual energy buy. Many of these measures were focused on load shifting as peak demand costs are a huge expense for the base. A 24-MW on-site generator has been installed, but it will require further upgrades in order to allow full control over load selection. In addition, expanded bike trails and bus service have made the base more accessible and reduced car emissions.

Base leadership understands that until they have energy storage, more on-site generation, and a working microgrid, risks to external energy supply will mean risks to the operational readiness of Fort Bliss. More collaborations are planned for the future. The base and local government are working together to connect bike lanes and bus services to facilitate non-car travel across the base line. To better manage waste, the city is interested in working with the base to develop a joint waste-to-energy plant of 40–45 MW.

Fort Bliss is dramatically reducing its energy intensity per square foot. They are on track to meet the federally mandated energy savings goals by 2015 and exceed them after that. They are beating their own goals for water conservation. The reason for this success is clear – open communication and collaboration of diverse stakeholders toward common goals.

The particular challenges and opportunities facing government actors in advancing energy resilience are discussed further in Chapter 7, "Government Energy Resilience: Policies and Programs."

Resilience Planning for Local Institutions

A variety of local institutions are recognizing an increasing threat to their energy system and, therefore, to their mission, from hospitals to universities and military bases. The local energy resilience planning process looks very much the same across various types of institutions. In all cases, the goal is to keep power on at specific facilities during grid outages. Long-term,

Figure 3.6 "Technical Resilience Navigator Modules."[5]

centralized control over these facilities can help make resilience planning successful, and resilience actions often include distributed energy resources co-located with the facilities they serve.

The Technical Resilience Navigator (a collaboration of the Federal Energy Management Program, Pacific Northwest National Laboratory, and the National Renewable Energy Laboratory) is an interactive resilience planning tool designed to help organizations assess their energy and water resilience gaps and identify actions that will help them to close those gaps. Its six steps are closely aligned with the steps laid out in this chapter (missing only performance evaluation as a final step), and each step is given a "module" that guides the user through completion.[5] The tool was developed with federal agencies in mind but is focused on site-level planning and is useful for any local energy resilience planning process.

The energy resilience perspective of local businesses, organizations, and governments is discussed in Chapter 6, "Energy Resilience at the Local Level."

Conclusion

The energy resilience planning process is all about successfully protecting people from the impacts and dangers associated with energy disruptions. For utilities, these people are also customers; for government, they are constituents; for a hospital, they may be patients. In each case, the functioning of the organization is vital to the functioning of society, and the health and safety of those they serve.

Whit Remer, the Sustainability & Resilience Officer for the City of Tampa, explained to us that "One lesson from Tampa's resilience planning is that it is crucial to engage those frontline communities that will be most impacted by climate change. We have set up a program to put a significant amount of funding into bringing frontline communities into the conversation to better understand their needs, their vulnerabilities, and their priorities. This community engagement is going to be embedded in every aspect of developing our resilience plans. We are going to talk to everyone, but those certain communities are critical."[6]

Starting with a risk assessment allows organizations to understand the size of the problem to be solved, and the planning process helps match this problem to solutions that will be both affective and practical to execute. There are many resources now available to help guide planners through this process. Using one of these guides as a starting point, an organization can then adapt it to fit its specific characteristics and mission.

The key for any organization is incremental progress. It is much better to set smaller, more achievable goals than to get so ambitious that goals can never be reached. In the next chapter, we will present a framework for an energy resilience maturity model and discuss how organizations can make progress toward greater energy resilience over time.

4

Resilience Metrics and Maturity Model

There is no "magic bullet" to combat our energy resilience challenges. We need to embrace a wide variety of energy solutions nationally and on a community level. Identifying, analyzing, implementing, and tracking all of these opportunities is a complex undertaking. Consequently, before taking any steps on this path, organizations need metrics that can compare potential investments, target program milestones, and measure success toward resilience goals. This chapter will discuss different ways to evaluate and track energy resilience initiatives, presenting a detailed outline for one of those methods.

The National Renewable Energy Laboratory states the challenge of measuring energy resilience quite simply:

> *"A resilience metric is used to quantify the ability of an energy system to prepare for and adapt to changing conditions, and withstand and recover rapidly from disruptions... No one definition or metric can be applied broadly; rather, the appropriate metric depends on goals, event context, hazards, scale, and perspective."*[1]

Although we would like to have a simple equation that gives us a resilience score comparing buildings, businesses, and utilities across the country, accuracy depends on specificity. Depending on the context, the risk factors, impacts of outages, and technical solutions will all be quite different. There cannot be a universal resilience metric the way we have universal reliability metrics like SAIDI and SAIFI.[2] That being said, several industry working groups, labs, think tanks, and government agencies are considering this problem and have devised a variety of approaches. The approaches include simple check lists, qualitative evaluation, and quantitative frameworks. As illustrated in the figures below, the levels of complexity and effort with these approaches can differ greatly. Figure 4.1, demonstrating an evaluation approach, focuses on the aspects of a resilience plan, presenting them alongside each other. Figure 4.2 is an example of a more quantitative approach which is more linear in nature and highly dependent upon detailed numerical scoring.[3]

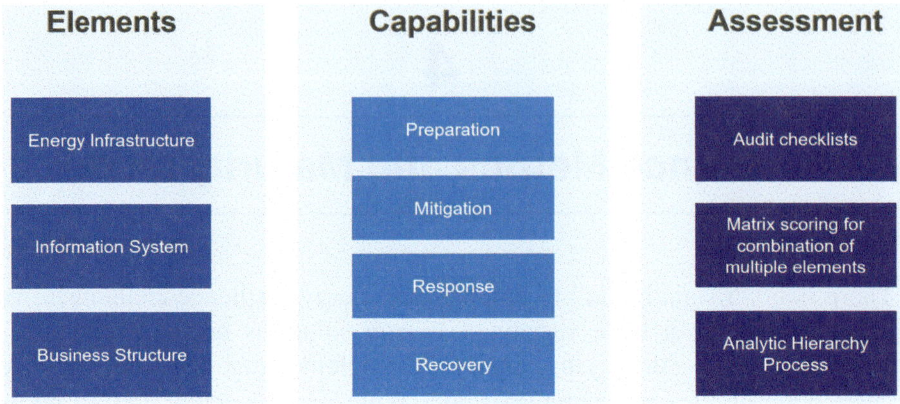

Figure 4.1 Qualitative resilience evaluation example.

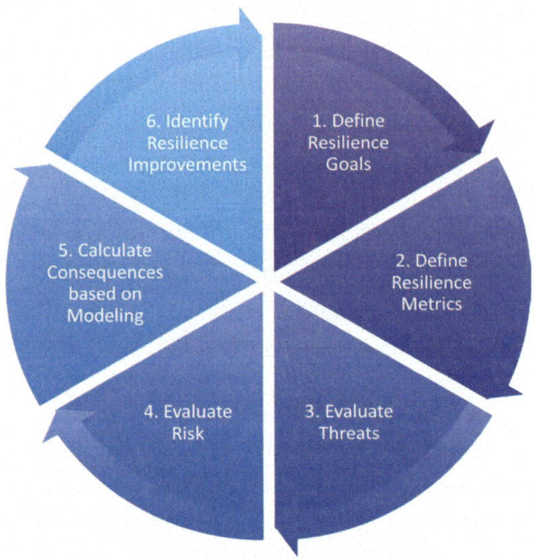

Figure 4.2 Quantitative performance assessment example.

Instead of being mutually exclusive, these approaches can be complimentary. In this chapter, we present three approaches that have very different use cases for evaluating resilience efforts. The first, a **maturity model approach**, is best for tracking an organization's success in addressing resilience programmatically. This approach, which indicates how far your organization has come in achieving its optimal spot on the spectrum of energy resilience, is also good for comparing resilience programs at different sites or one organization against another. The second is the **resilience value approach** – a simple risk identification and avoidance metric that is best used to compare various resilience investment options. Finally, there is the **customized metric approach**. This method is complex and limited in application and represents the only way to quantify the impacts of a resilience program against a specific set of threats.

Not every organization will need or want to adopt all three of these approaches, but understanding them better will allow you to invest in the most helpful evaluation method. Across all methodologies, however, the key to success is to gather the right stakeholders to agree on 1) the risks to your organization's energy resilience, 2) what assets/resources need to be resilient, and 3) how your organization defines "resilience."

Before developing your evaluation framework, make sure your organization has the data tracking and internal expertise to measure, monitor, and interpret the data. Erik Svanholm of S&C Electric put it bluntly when he said, "A critical question resilience planners need to ask is 'Do we have the internal expertise in our organization to properly identify and evaluate these solutions?' Do NOT assume that the answer is yes!"[4] As energy resilience is a relatively new area of expertise, you may need some internal training or even external consulting to support the initial phase of your project.

The Maturity Model Approach

The simplest approach to measuring resilience is to identify the programmatic elements that lead to greater resilience (such as an emergency response plan, backup generation, or specialized training) and to measure your organization's progress in getting those programmatic elements in place. This "maturity model" approach inherently acknowledges the difficulty of assigning values to a complex and changing system. It is also helpful for comparing organizations against each other – even when their resilience threats and response capabilities differ.

A common format for a maturity model is to list levels of maturity (ranging, for example, from "No Program" to "Best in Class") and then list milestones or characteristics of operations at each level. A simpler format is to list critical activities for achieving resilience and classify them as *Not started*, *In progress*, or *Complete*. You can even assign points to these positions to create a "resilience score" if desired. That can be useful in comparing sites against each other or for rudimentary performance comparison over time.

A realistic understanding of your current status, using the maturity model, will help determine the goals of your organization. Not every aspect of your program, for example, needs to be best-in-class. Often, it makes more sense to target areas of highest vulnerability or greatest potential.

Marisa Aho, Chief Resilience Officer for the City of Houston, proposes beginning with a resilience assessment. "You have to start by identifying the issues that could affect your resilience and how well prepared you are to deal with them today. Demonstrating co-benefits of action across multiple shocks and stress is very helpful – it builds the business case. It shows cascading impacts that one disaster can have on multiple systems and populations."

The U.S. Department of Homeland Security (DHS) utilizes maturity models in their in-depth Resilience Framework document.[5] In the process of developing this framework, DHS created a "Resilience readiness assessment planning score" for each of their sites. The agency determined what it would need across a range of operational functions to consider a site "resilient" and then gathered that data through on-site visits and responses to questionnaires – yielding a score by which to compare sites.

This process works best when there is agreement around the organization's resilience goals. In an interview, Crystall Merlino, Director of Resilience and Energy Management at DHS, talked about the importance of engaging stakeholders. She explained that asking questions of the organization helped to bring in stakeholders and get them engaged.[6] For instance, asking your stakeholders to identify what is mission critical and what is in need of greater resilience can help you to bring additional parties to the table. This kind of engagement is critical to identifying the milestones or success factors that are going to be used in the model itself.

The scorecard shown in Figure 4.3 from DHS's framework shows how sites are graded across several performance areas. While score cards like this do allow for multiple site comparison, they may ignore the relative importance of each line item for a specific site. Director Merlino expounded on DHS's use of this scorecard stating "The tool doesn't tell us how resilient we are but what our vulnerability is. Reducing your exposure to vulnerability is improving your resilience. We are now trying to piece it together through prioritizing our funding -- working with our financial group, property team and other stakeholders to highlight our most vulnerable and critical facilities."

APPENDIX D: SAMPLE AGENCY SCORECARD

Metric Score		
Not Started / No	In progress	Complete / Yes
1	2	3

Step	Metric Category	Score: Information Communication Technology	Score: Energy and Water	Score: Facilities	Score: Transportation
1	Preparedness/Planning				
1.1	Stakeholder Identification	3	3	3	3
1.2	Critical Mission Identification	3	3	2	3
1.3	Criticality Assessment	3	3	2	3
1.4	Risk Assessment	3	3	3	3
1.5	Cyber Plan	3	3	3	1
1.6	Information Sharing	3	3	3	1
1.7	Business Continuity Plan	3	3	3	3
2	Construction Mitigation Measures				
2.1	Natural hazards	3	3	3	3
2.2	Standoff distance	3	3	2	3
3	Alternate Site Mitigation Measures				
3.1	Mitigation with Alternative Sites	3	3	3	3
4	Resource Mitigation Measures (Redundant Sources)				
4.1	Electric	3	3	3	3
4.2	Natural Gas	n/a	3	3	n/a
4.3	Communications & IT	1	3	3	n/a
4.4	Transportation	n/a	3	3	2
4.5	Water	n/a	3	3	n/a
4.6	Wastewater	n/a	3	3	n/a
5	Response Capabilities Onsite				
5.1	Incident command capability	3	3	3	3
6	Response Capabilities Offsite				
6.1	First responders interaction (mutual aid)	3	3	3	3
7	Recovery Restoration Agreements in Place				
7.1	Agreements	3	3	3	n/a
8	Resource Recovery Plans				
8.1	Recovery Plans	n/a	3	3	3
	Total Score	2.9	3.0	2.9	2.7

Figure 4.3 DHS site resilience checklist example from their Resilience Framework plan.

Maturity models, while useful in highlighting a broad range of program elements, also have considerable shortcomings. These models are unhelpful when it comes to prioritizing investment options. They also provide very low resolution around any progress being made, since they are based on stage gate accomplishments rather than incremental improvements. Finally, maturity models only work well if there is a high degree of confidence in your understanding of the right elements to include. Further down, we offer suggestions on how to structure that evaluation.

The Resilience Value Approach

When an organization is less interested in comparing sites against each other or against a plan and more interested in comparing the value of different investment options, the resilience value approach often makes more sense. A resilience value can be a combination of a variety of value streams such as loss of utility revenue, cost of damage to the grid, loss of assets, or business interruption costs. Attaching monetary value to these streams allows your organization to understand the resilience value of certain actions. This valuation should also consider all other potential value streams, such as the following:

- Reduced insurance rates
- Reduced mortgage rates
- Government incentives
- Grid services value
- Resilience payments from site host[7]

STEM Energy Superintelligence, an energy storage firm, developed a white paper titled "Designing a Policy Roadmap to a Clean Resilient Grid," which outlines an approach that involves creating resilience values.[8] The approach begins with three input categories: outage cost, outage risk, and level of resilience provided. These three inputs can establish a "base resilience value" which estimates the financial losses prevented by a resilience solution. This solution does not do much to assess your system's level of resilience but is useful for comparing potential resilience investments and demonstrating the return on investment for each one.

The first step in the resilience value approach is to determine expected outage costs across the most common facility and operation types. This cost can be segmented into the fixed costs associated with an instantaneous loss of power and variable costs associated with the duration of the sustained outage. What happens if a storm knocks out power for two days? What happens if on-site generators are not properly maintained? Whatever the risk is, you need to understand what the likely impact is going to be to your operations. Most of these impacts are likely to relate to suspension of operations; so you need to understand the full costs should your facility shut down.

There could be other financial impacts to consider as well. If there are ongoing tests that would be disrupted, that is an additional cost beyond the hours lost during an outage. If refrigerated product could spoil during a long outage, that loss of inventory needs to be factored in.

Next, outage risk is estimated by forecasting the expected outage frequency and duration. The risk is estimated for a set time horizon (e.g., 20 years). The exact nature of the risk depends on the type of facility. While we commonly think about storms, an aging grid, and cyberattacks as major threats, a given site may have other, more specific, threats. Supply chain can be a threat. Extreme heat, cold, floods, or drought can be a threat. Employee incompetence or lack of training could be a threat. And, as COVID-19 has demonstrated, community disease can also be a threat to resilience.

Damir Novosel, President of Quanta Technology, wrote a white paper on resilience metrics and offers this advice on focus: "Everyone always focuses on weather. Now we think about cyber. But system design is a major issue. You have to address them all. If you only address weather, you will ignore the others. There are usually triggering events."[9] Convening a broad range of the right stakeholders should give you a holistic look at potential threats.

For any given project under consideration, you should analyze how likely it is to mitigate energy loss from all potential sources or a specific source.

Each investment is then given a Resilience Service Level score, which is a combination of fixed and variable impacts based on how much an investment is likely to mitigate risk. These scores can be weighted based on the priorities of the organization. Some options will be better for achieving energy self-sufficiency, while others will focus on mitigating outage impacts or addressing needs of underserved communities. Finally, by comparing the costs of doing nothing versus initiating in a resilience project, decision makers can determine what the estimated rate of return would be on resilience investments.

A similar but slightly more complex approach can be found in the Grid Modernization Laboratory Consortium's (GMLC) April 2020 paper, "Grid Modernization: Metrics Analysis – Resilience."[10] This approach, called multi-criteria decision analysis (MCDA) is a two-stage process, visualized in Figure 4.4. The first stage uses stakeholder engagement and option valuation to develop a suite of resilience options (similar to a maturity model). The second portion applies a risk-focused cost–benefit analysis to each option. As these options are implemented, real-world results allow planners to revise their evaluation of each approach and reconsider which alternatives to pursue. This model, in application, is more focused on the needs of electric utilities with their highly complex systems and would be harder to employ for facility owners.

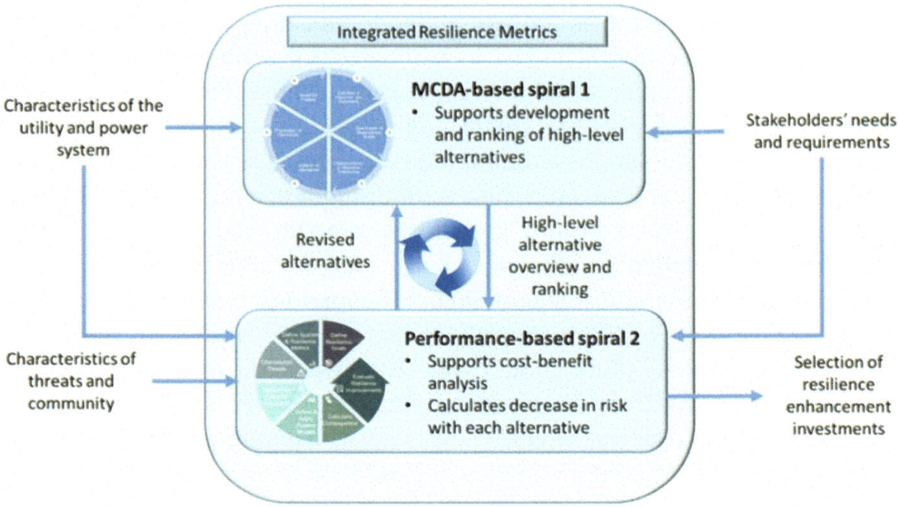

Figure 4.4 Schematic of the GMLC's integrated resilience approach.[11]

Inputs	Approach	Outputs
• Enterprise overall planning and management processes. • Characteristics of electricity distribution assets. • Electric distribution system alternatives. • Performance-based resilience enhancement options.	MCDA	• Characterization of the system-level distribution of preparedness, mitigation, response, and recovery capabilities. • Resilience index value of the distribution system, ranging between 0 and 100. • Ranking of electric distribution system enhancement alternatives.
• Characteristics of threats and community. • Consequence categories. • MCDA Alternatives and their RI values.	Performance-based	• Outage impact. • Recovery costs. • Community impact. • Resilience enhancement options.

Figure 4.5 GMLC's inputs and outputs table for the two analytic spirals.

The Customized Metric Approach

If you really need a numerical scoring system to track changes in the energy resilience of your facility or organization, you will need to acknowledge certain limitations. As we have discussed, resilience challenges differ across organization type and geography. When you introduce different risks and different capabilities to mitigate risks, it becomes less and less practical to measure performance with the same formulas. However, if you are able to create a metric taking these unique factors into account, you can gain a much more granular understanding of resilience performance over time.

One of the most comprehensive treatments of resilience metrics is described in IEEE's report "Resilience Framework, Methods, and Metrics for the Electricity Sector."[12] Here, the authors make the case for the need for resilience metrics but also acknowledge that it is not possible to create a simple set of metrics that will work across all organizations or geographies. They propose identifying individual parameters/events that need to be tracked and creating system-specific metrics for each. These metrics, properly weighted, can then be used to evaluate investment decisions. The customized nature of the metrics makes them hard to compare across utilities or regions. More likely, the best use of these metrics is to track performance improvements over time within an organization.

The framework itself is based on an "all-hazards" approach for assessing and developing a program, with five main focus areas: prevention, protection, mitigation, response, and recovery. The goal is to minimize the likelihood and impacts of a disruptive event and provide the right guidance and resources to respond and recover effectively and efficiently when an incident occurs.

The IEEE report is directed at electric utilities, and it asserts that the first step to making the grid more resilient is to effectively address preconditions that cause grid stress, such as the following:

- Local grid congestions
- Insufficient generation resources
- Power quality challenges
- Insufficient planning for transmission and distribution needs
- Insufficient data about grid operations, developing load needs, or external stressors
- Regulatory uncertainty resulting in insufficient investment
- Human error

An example of this all-hazards approach can be found at Southern California Edison, which used the following areas of focus to create its resilience strategy:

- Protection: Protecting the company from physical, social, cyber, and financial threats through asset hardening, barriers, and specialized equipment to strengthen critical assets.
- Prevention: Preventing disruptive events that could negatively impact the company, customers, employees, and/or infrastructure through

intelligence and information-sharing that drives tactical decisions to avoid or stop disruptive events.

- Mitigation: Mitigating the impacts of an incident by developing strategies that reduce the company's vulnerabilities, risks, and the loss of resources, life, or infrastructure.
- Response: Responding to all incidents with a uniform approach, consistent with those used by the emergency management community, public agencies, and first responders.
- Recovery: Recovering using established plans and procedures to quickly get the company and the communities it serves back to a state of normalcy while ensuring appropriate corrective actions.

Across all the approaches we have seen to resilience planning, the basic concept is the same:

- Understand today's threats to your organization's resilience – either through a qualitative assessment or a quantitative scoring system.
- Attempt to quantify that risk (in terms of financial or other impacts).
- Identify resilience improvement options.
- Evaluate the relative impact of those options against each other and against doing nothing – weighting their benefits through the lens of your organization's specific priorities.
- Track performance – again through qualitative or quantitative assessment.
- Unfortunately, there is not one simple system that can be utilized for quantitative scoring across all organizations. But any analysis needs to incorporate data on preparation, perseverance, adaptation, and recovery. Investments in preparation and perseverance may result in input data that makes adaptation and recovery scores low. Likewise, building a system that fails easily but is much better at recovery would create different looking scores with the same methodology. What is clear is that resilience planning requires qualitative evaluation of the most granular quantitative information you can *actually collect*. Be realistic on the level of detail you can expect to receive.

Getting Started with the Maturity Model

The best place to start is usually with a maturity model. The exercise of creating this document with your stakeholders will advance your thinking around what needs to be done to move you along the spectrum of energy resilience. What is more, the conversations around the model will tell you whether more detailed metrics or scoring are necessary. If you find yourself asking questions such as, "How will we *know* when a project is complete?", you might want to put extra effort into resilience valuation or a custom metric.

Breaking successful energy programs down into their key components will help you assess progress in each individual area. Results are certain to be mixed, but even uneven progress in multiple areas will result in an overall shift toward a more resilient energy system for your organization. While you can use whatever categories make sense to you, we believe that a good starting point would include the following:

- Goals
- Strategy
- Project financing
- People and maintenance
- Technology solutions
- Supply chain
- Information management

The model examples later in this book sort status into five levels of program maturity – from "no program" through "best-in-class." Here is a brief description of each of these levels for the purpose of this analysis:

- **No Program** – Energy is viewed as a fixed cost and there are no initiatives to address potential threats.
- **Reactive Program** – Energy resilience is valued, with an organization willing to consider efforts on this front. However, there is no effort made to seek out these opportunities and there are no clear guidelines on funding such efforts.
- **Managed Program** – This program has guidelines around energy investments and may even have goals for managing energy as a resource. Here, the focus is typically on cost savings, rather than evaluating projects on the broader criteria of security and risk avoidance. This is the kind of program we find most often in corporations because it maximizes near-term economic outcomes.
- **Proactive Program** – This program recognizes that pursuit of energy resilience has long-term benefits that justify major investments. In this situation, organizations work to proactively identify energy resilience challenges and address them with planning and investment. This is where we see many electric utilities today but not uniformly.

- **Best-in-Class Program** – This category is reserved for efforts that meet the criteria of a proactive program, with action steps implemented in a holistic, integrated manner. This category requires considerable investment of time and money, featuring initiatives that address all aspects of energy resilience and leverage the strengths of the organization itself.

Conclusion

At the end of the chapters on local entities, electric utilities, and government programs, we will provide a sample maturity model. In summary, the purpose of this model is to help your organization diagnose resilience across multiple areas so that you can make informed planning decisions. The goal is not to try to wrestle your organization into the best-in-class program in every category. Achieving that is close to impossible and likely a wasted effort for most organizations. Instead, this model hits the highpoints, ensuring that your organization is addressing each important section of the road to energy resilience.

Part 3

Energy Resilience at Every Level

5

Energy Resilience at the Grid Level

When outside forces threaten our energy system, the grid itself is the first responder. The best strategies for resisting these threats involve strengthening the grid itself. Resisting physical impacts, foiling bad actors, and reconfiguring on the fly to contain damage are all ways that the electric companies try to support local and national resilience. Once the grid goes down, local solutions spring to the fore, but most of these are stop-gap measures. It is up to grid operators to get power flowing again in short order to ensure that a bad situation does not become catastrophic.

In this chapter, we will discuss current challenges to grid resilience, some best practices for utility resilience planning, specific solutions utilities are leveraging to improve resilience, and finally, how to incorporate these kinds of solutions into rate cases.

The Changing Needs of the Grid

As noted in Chapter 1, the threats to our energy grid are increasing. Our energy system is aging, extreme weather is worsening, and cyber-attacks are becoming more prevalent. If nothing else were changing, energy companies would still be facing a daunting task to improve the resilience of the grid. There are, however, additional changes that affect the way we use and generate electricity, and this evolving landscape will make a resilient grid that is much more vital. The good news is that these changes also bring some opportunities to support that resilience – but only when communities plan properly.

The Increasing Cost of Outages

With each passing year, we are asking the electricity grid to do more for us – and the strain is showing. Our reliance on electricity for nearly every function of society has increased the impact of grid outages. A 2021 S&C/ Frost & Sullivan report showed that for the fourth consecutive year, power reliability was not improving.[1] Among the participating companies, 44% indicated that they lost power at least once a month in 2020 compared to 21% that experienced this problem in 2019. Roughly a quarter of survey respondents (22%) said they lost at least $100,000 per typical outage. Of these companies, 80% indicated that they experience outages at least monthly, which means they stand to lose at least $1.2 million annually because of power outages.

The S&C study was focused on typical short-term power outages, and not major outages like those that follow a severe storm. Every case study of major outages in the past 20 years has shown a dramatic economic impact on local communities.

The Potential of Distributed Energy Resources

The falling prices of solar generation and energy storage, combined with programs supporting energy efficiency and demand response, have encouraged the proliferation of distributed energy resources (DERs), such as rooftop solar. Given the right signals and incentives, DERs can add to the resilience of the grid during times of stress.

Local generation and energy efficiency programs reduce the loads on local grid elements, allowing energy companies to extend the lifespan of their equipment and operate with slightly more capacity headroom. Energy storage and demand response capabilities allow for modulation of loads during times of congestion, again reducing the need for costly wires investments but also giving the utility some short-term options to reduce demand when an external threat drives demand up or limits supply.

Adding renewable generation such as distributed wind and solar resources helps to diversify an electric company's energy supply and helps protect it from threats like fuel supply chain and water shortage.

None of these measures will simply happen, however. Incorporating DERs is not as simple as plugging in a solar panel. It is important to remember that the grid was not designed for this. For the last century, the grid has operated as a one-way power delivery system. Power plants were located at one end of the grid, with customers at the other. Today, DERs enable power to enter the grid at a variety of locations, which is changing the architecture of the grid itself.

The grid is changing from a one-way power delivery system to a distributed electricity network, facilitating generation and loads in any direction. When left unchecked, this shift in power flow can pose significant risks for existing equipment on the grid. Before utilities can fully embrace DERs to help support generation needs and support customer demand, other aspects of the grid and overall utility operations need to be updated to better accommodate this change. When done correctly, the installation of DERs can improve a grid's reliance and reliability.

The Demands of Electrification

New loads are going to increase demand on the grid but could also represent an opportunity to strengthen it. Consider the growing popularity of electric vehicles (EVs). Bloomberg New Energy Finance estimates that battery density will continue rising at 4%–5% per year, and EVs will achieve cost-parity (without subsidies) with internal combustion vehicles by the mid-2020s.[2] Adding these vehicles to a grid increasingly powered by clean sources will be a major win in the battle against climate change as, in the U.S., roughly 28% of carbon emissions come from the transportation sector.[3]

At the same time, electrifying our transportation sector will mean that sustained electricity outages will have an even broader impact than they do today. Assuming vehicle batteries reach range parity with internal combustion vehicles (which they are very close to doing today), taking stored fuel out of the equation means that any grid outage will put electric vehicles on the clock. Distributed energy solutions and microgrids at charging stations could help meet this challenge, but those solutions will need to be put in place if electrification fails to increase our exposure to grid disruptions.

For its part, the electric utility industry seems on board and ready to invest in the infrastructure needed to make the transition to EVs work. In early 2021, after President Biden released an executive order directing U.S. federal agencies to transition their vehicle fleets to EVs, the Edison Electric Institute (the trade association representing investor-owned utilities) issued a press release stating, "...we are particularly pleased by the administration's executive order to electrify the federal fleet. The transportation sector is the largest domestic source of carbon emissions. By accelerating transportation electrification and increasing the number of electric vehicles in the federal fleet and on U.S. roads, we can leverage the already ongoing emission reductions in the power sector to meet economy-wide carbon reduction goals."[4]

While the proliferation of passenger EVs may not immediately stress the grid, clustered charging of fleet vehicles is going to be a real, near-term challenge for utilities unless they develop their ability to reroute power and manage higher demand peaks. In 2020, S&C Electric Co. partnered with several Canadian distribution companies to apply the data from their feeders (including distribution system models) to make an assessment on how the addition of EV charging would impact those feeders. The findings of the study indicated that high but realistic future penetration levels of EV charging would likely result in dangerous voltage drops at feeders loaded at 50% or more of their maximum capacity, assuming no other mitigations were applied.[5]

On the heels of vehicle electrification, we are already witnessing a push to electrify building heat and hot water. This is a worthy goal in the fight against climate change, but this movement will add additional load burden electricity grid.

The Resilient Grid is a Smart Grid

To be resilient, our grid will need to get smarter, more distributed, and more reconfigurable in real time. Regulators and electricity companies must come together to invest in a grid capable of digital, multi-directional flows of information and energy. This shift to a "smart grid," leveraging the latest technological innovations, can provide far-reaching benefits including lower electricity bills and a cleaner energy mix.[6] A well-designed smart grid provides additional resilience benefits that include redundancy, reduced outage times, faster identification of grid failure, and the ability to shape demand and reroute power flow to minimize impacts.

Erik Svanholm of S&C Electric Co is optimistic about energy companies' willingness to embrace these new solutions. "Awareness of proven technologies and solutions is becoming more common. The technical opportunities are more salient. At the same time, the spectacular regional outages are going up in frequency and visibility. The need for utilities to be seen as taking proactive action is increasing. News is instantaneous now and that drives up accountability for utilities because it makes people's outage experience more visible nationally. People are tweeting photos of themselves shivering in their living room and that creates a media cycle that puts a harsher spotlight on the utilities."[7]

The hardest part of addressing resilience on the grid is that the grid is simply everywhere. As Bennet Chabot of Pacific Gas & Electric told us, "It is a ubiquitous piece of everyone's life that no one thinks about. We haven't demanded that people understand energy infrastructure because they haven't had to."[8] Now, politicians, regulators, and customers need to start thinking about the changing needs of the grid, and how they can advance this transition. In the meantime, utilities will need to conduct resilience planning to do the best with what they have.

Best Practices for Utility Resilience Planning

Thinking about the resilience of the grid is not new for electric companies. In their foci on safety, reliability, and long-term investments, utilities have always taken action to keep the power flowing during major events and get it up and running quickly when the grid goes down. Because of the rising challenges discussed in this book, electric companies now need a more aggressive and proactive resilience strategy to keep up. Creating and implementing such a strategy requires the adoption of some proven best practices.

Understand the scope of what you want to accomplish

The spectrum of energy resilience looks different for energy companies than it does for end users. Each electric company may have a different optimal location along the spectrum. The first step of any resilience effort is to understand what your operating definition of "resilience" is, what scope of activities you will address, and how you will measure your success. Only with this clarity of scope and purpose will you be able to prioritize your investments and understand where you are being successful.

Adrienne Lotto, the Chief Risk and Resilience Officer at New York Power Authority, made this point: "A major challenge is simply the breadth of the problem. Resilience can mean different things to different people depending on where they sit in the organization. This is great because it allows everyone to get behind resilience. But in other ways it can make prioritization and resources a challenge. If you took any bucket that we deal with on resilience, we could easily spend our entire budget on it. But you have to start somewhere – make a beginning and then build on it."[9]

Once you have gathered the right decision-makers within your organization to develop a resilience plan, the first steps should be the following:

1. creating an organizational definition of resilience;
2. deciding what metrics will be used to track your progress on resilience;

3. establishing an organization-wide goal around improving energy resilience, based on your desired end point on the spectrum and the major milestones to reach that goal.

Recruit the right skill sets

It would be a mistake to assume that, because you are in the business of operating a reliable electricity network, you have all the in-house skills necessary to make that network resilient. A focus on resilience is going to require an ability to look at threats in new ways and implement solutions that would not necessarily make sense for a business focused only on reliability.

Erik Svanholm of S&C Electric summarized this in an interview: "There are technologies and techniques that can be applied – but [utilities] know that they are not up to speed on what these should be and they don't have the bandwidth to stop running their system to go learn about what to do. Few utilities have chosen to learn about the resilience world so they can evaluate investments. They always treat it as something to deal with in a few years. That's shifting."[10]

In addition to traditional skill sets like system planning and repair, any utility looking to make major improvement in the resilience of their system should take stock internally to understand if they have in-depth expertise in the areas of the following:

- Climate change impacts
- Cybersecurity
- Market transformation and electrification
- Distributed energy system design
- Energy storage
- Emergency planning (including coordination with first responders)

These skills can be developed internally through hiring and professional development. There are also consultants who have specific expertise in these areas who can be brought on to identify gaps and help establish programmatic roadmaps.

Understand your Threats

One you have your goals and metrics laid out and you have the right skill sets in place, it is important to determine the variety of threats faced by your utility. Threats can be defined as anything that can expose a vulnerability and damage, destroy, or disrupt the power system. Threats can be natural, human caused, or technological, and they are not typically within the operator's control. They can include wildfires, hurricanes, storm surges, cyber-attacks, and more. These threats differ based on geography, utility size, generation mix, and load composition. A large, well-known utility in Texas, for example, is going to face a dramatically different threat landscape than a small municipal utility in Ontario.

Any threat consists of *exposure* and *impact*. Exposure is how likely it is that a particular threat will adversely affect your system. Impact tells you how much trouble that threat will cause once it happens. Ron Critelli of Florida Power & Light proposes focusing on exposures first and then letting impact dictate how much effort you will target at a particular threat. In an interview, Mr. Critelli talked about controlling for an element first when designing a transmission line – whether that be wind, ice loading, or even wildfire. Once you look at your exposure, you can then evaluate outliers and decide if you need to design for those. Once you have your hands around what you are going to try to control for, you can then think about how to deploy your available budget for altering the distribution system.[11]

The "Power Sector Resilience Planning Guide," written by experts at the National Renewable Energy Laboratory and the United State Agency for International Development, is an excellent resource for understanding threats to the energy system.[12]

Perform Targeted Stakeholder Engagement

Stakeholder engagement is critical for utility resilience planning. You cannot understand the needs and capabilities of your network without connecting with the stakeholders who operate, maintain, and are served by that network. Selection of stakeholders will depend on the focus of your resilience efforts but may include the following:

- the line crews, who know the realities on the ground and what can and cannot be accomplished;
- network planning leadership, who understands where demand needs are changing and what new resources are coming onto the network;
- the finance team, who understands how financial decisions are made and can help develop proposals that will make it into a rate case;
- your customers, who can help you to focus your resilience efforts where they will do the most good.

Different types of end users will have different resilience needs. If your line consists largely of residential customers, you will be less concerned about short-term outages and more focused on keeping critical loads like first responders up and running. Those critical loads may go well beyond just police and fire departments. Water and natural gas infrastructure in your service territory are sure to be critical loads as are grocery stores, pharmacies, and even morgues.

If you are dealing with more commercial and industrial customers, you need to understand their specific needs and capabilities. For example, one customer may have on-site refrigeration that can last for 48 hours on backup generation. Another site may have on-site generation or a microgrid allowing them to tolerate a power outage for longer.

Prioritize System Components

Once you understand your exposures and impacts, you can start to prioritize loads and, therefore, which parts of your system to protect first. You can anticipate which aspect of the system is more important to protect from damage and apply your primary focus there. It may even be the case that letting other parts of the system fail is part of a larger strategy to protect the most critical elements.

This is exactly the strategy that Florida Power & Light used when tackling the impact of high winds on their lines. FP&L never imagined that they were going to be a failure or outage-free. Instead, they looked at what drove restoration time. For FP&L, that was replacing poles. They then designed a system that would severely curtail the number of pole failures. All of the other failures such as cross arms, insulators, and wires were quicker to address than pole failures which in the end allowed for faster restoration times. FP&L knew their system would go down. They designed it to be down for as short a time as possible.[13]

Make Resilience an Integral Performance Factor

Many utilities have thought about resilience as an expansion of their reliability efforts. While there are differences in the drivers and solutions for resilience and reliability, many of the solutions for one will bear fruit for the other. The challenge is that utilities are already accustomed to tracking reliability and using it as a system performance factor. Building a resilience metric for grid operation is a new and much more challenging task.

Whatever type of performance tracking you ultimately decide upon, it is important that resilience efforts are built into the DNA of each project an energy company undertakes. Incorporating resilience measures into the design of new projects will be vastly easier than later attempting to add resilience measures to existing infrastructure. This point was stated succinctly by Sharla Artz, Director of Security and Resilience Policy at Xcel Energy, in "Build it in versus bolting it on."

What Can Utilities do to Improve Resilience?

There are a variety of concrete actions that a utility can include in their strategic plans to affect resilience. Broadly, these actions fall into three categories:

- Prevention: Actions you take ahead of time to reduce exposure to threats.
- Response: Actions you take when threats like cyber-attacks or extreme weather are stressing the grid.
- Recovery: Actions you take after the grid is impacted to get things back to normal operation as quickly as possible.
- It is difficult to categorize some of these solutions below because many of them could fit into more than one of these categories. The scope and design of these solutions may be different whether you are attempting to implement them for resilience versus reliability, recovery versus prevention, etc. It is also important to remember that these solutions do not live in silos. Big data analytics can inform vegetation management. A grid segmentation plan must be informed by other infrastructure buildout plans. When implementing any of these solutions, think both about how your projects would contribute to all three categories of resilience and how they will inform or support each other. Figure 5.1 demonstrates some of the most common energy resilience solutions for the grid.

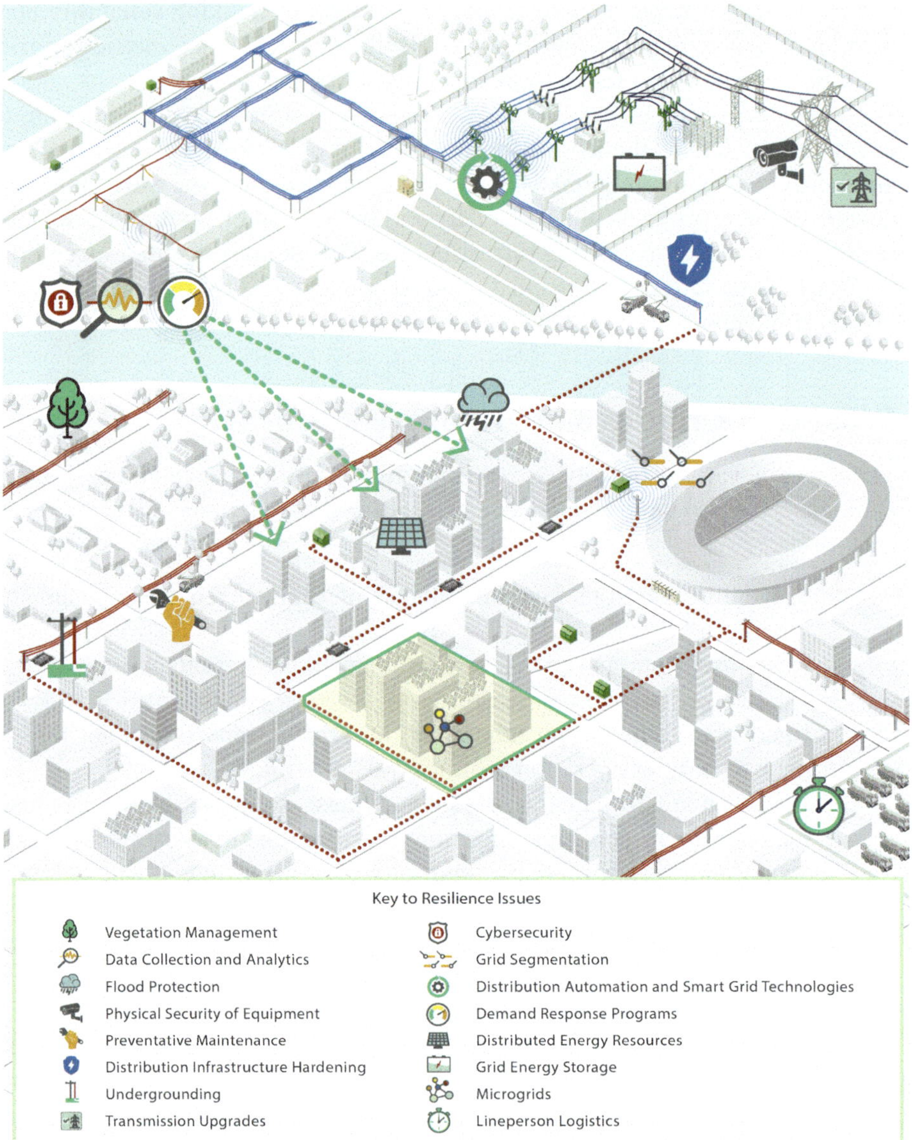

Figure 5.1 "Electric Grid Resilience Solutions." S&C Electric Company, 2021.

Vegetation Management

Trees are a leading cause of electric outages in the United States with an annual economic impact exceeding $100 billion.[14] While not typically considered cutting edge or innovative, vegetation management is the most important step many energy companies can take to improve both reliability and resilience. A tree strike anywhere along a distribution line takes down the entire line – impacting both customers and the utility costs. Vegetation management is also critically important in areas where utility equipment may be the cause of wildfires.

The biggest recent improvement in vegetation management has been the use of satellite data, drones, and artificial intelligence to understand where trees may pose a risk to power lines. Depending on the context, the use of this kind of technology for vegetation management may have the highest benefit-to-cost ratio of any resilience investment.

Data Collection and Analytics

The old adage of "you can't manage what you don't measure" has been recited in energy industry presentations for decades. Recently, the scope of what needs to be "managed" in the energy industry has been increasing. A February 2021 report from the Department of Energy's Electrical Advisory Committee stated that data sources are increasing due to smart meters for field measurements and electric vehicle charging, radar for weather measurements and asset monitoring, and embedded sensors for condition-based observations.[15] There are now grid-edge sensors in equipment such as smart switchgear and AMI meters, and they are telling us more each year about how the grid is performing, how we can best balance loads with supply, when and where to perform equipment maintenance, and the effective lifespan of each piece of equipment in its specific application.

Bennett Chabot, Grid Innovation Product Manager at Pacific Gas & Electric, shared his hopes for future data use: "Data collection/models/algorithms let us know when risks are actually happening. If you could see the gust of wind coming and you only shut down lines while the wind was blowing, that would be a magical capability. The goal is getting information about the system and doing the right things with it on the right timescale."[16]

A good example of this in practice comes from Toronto Hydro, which is using transformer monitor technology to see which transformers are overloaded and calculate equipment end of life. In turn, this allows the company to upgrade equipment before it fails and helps to improve service reliability.[17]

Today, data is getting so rich and granular that we need to be careful to filter out what we do not care about and using big data analytics to identify important trends. Big data analytics can be defined as the process of uncovering trends, patterns, and correlations in large amounts of raw data to help make data-informed decisions. The skills and software to do this effectively are in no way a given in the utility industry. To fully inform your grid decisions, you must have both data from the grid and the capability to crunch that data to look for trends and prediction capability.

Flood Protection

Extreme rainfall events and sea level rise are both contributing to increased flooding, which in turn poses a significant resilience risk to the energy grid. Forty-one million Americans live in an area at risk of flooding from rivers and streams while another 8.6 million Americans live in areas susceptible to coastal flooding.[18] There are additional concerns about flash flooding and urban flooding as we have seen in cities like New York after heavy rainfall.

Electric utilities preparing for flood events need to think about what equipment might be affected by a flood and what flooding of that equipment would do to the grid at large. It is ideal, but not always practical, to put utility equipment above or outside of flood-prone areas. Another solution might be to employ fully submersible switchgear that can operate even if flooded. Keep in mind that the damage from flooding may not be purely from water but also from debris that is carried by that water or even the release of harmful chemicals.

Physical Security of Equipment

Physical security of grid equipment can involve a variety of approaches from perimeter fencing to monitoring and hardening locked cases for equipment. Although it is important to protect equipment from accidental damage (like being struck by a tree or automobile), the resilience aspect of security requires protecting the grid from bad actors. Particularly when combined with cyber-attacks, physical attacks at one location can have wide ranging impacts.

For the purposes of energy resilience, utilities must recognize that the most important physical areas to control are those that would give perpetrators access to networked digital equipment. In addition to maintaining physical barriers, AI-enhanced video surveillance, and locks, energy companies should consider and control how much access an individual might have to the utility digital network if they were to penetrate those layers of physical security.

Preventative Maintenance

While preventative maintenance may seem like only a reliability solution, it can be relevant or even critical to resilience as well. Considering what the needs of the grid are on a blue-sky day and running a preventative maintenance program to meet those needs will improve reliability. What happens, though, when the grid is stressed? Is the equipment in the field ready to handle impacts like heat, water, and wind? Can existing switchgear handle increased loading from rerouted power when other parts of the grid fail?

Preventative maintenance is a great example of how a traditional utility activity could be altered considerably by "building in" resilience in the data collection and design of the program. A preventative maintenance program will have significant overlap with vegetation management and with the use of big data analytics. This is best illustrated in Japan, where utilities like Tokyo Electric Power Company Holdings Inc. (TEPCO) are using new data collection procedures and big data analytics to better predict the maintenance needs of each individual piece of technology on their system.

Distribution Infrastructure Hardening

System hardening is about reducing your system's exposure to potential damage. A concrete poll is harder to knock down than a wooden one. The addition of guy wires will increase structure strength without the need for full pole replacement. Florida Power & Light will have their 20-year grid hardening effort completed by 2025. Internal statistics have shown 41% better performance in day-to-day reliability from hardened vs. non-hardened feeders.[19]

In this area, it is wise to remember the best practice about prioritizing system components. It is impossible to create a system so robust that it cannot be broken. It is better to harden the most difficult things to fix, allowing other aspects of the system to fail if need be. As part of the effort mentioned above, Florida Power & Light focused on poll strength, knowing that power lines could be added back to polls more easily than replacing the polls themselves. In some areas where even poll hardening seemed insufficient, FP&L utilized undergrounded lines – replacing fuses with reclosers to improve protection on those laterals.

Undergrounding

Placing utility lines underground eliminates their susceptibility to wind, ice, and lightning damage. It also rids neighborhoods of unsightly power lines. Unfortunately, burying lines is often prohibitively expensive and can make faults on damaged lines more difficult to find and repair quickly. That said, several utilities (in dense urban areas or locations prone to hurricanes) have decided that the benefits outweigh the costs. Sometimes, entire neighborhoods are undergrounded (particularly easy in the case of new housing developments). In other cases, undergrounding has been limited to specific laterals serving critical infrastructure or selected backbone circuits.

In the U.S., both Florida and Virginia have fundamentally changed their rate case process to allow utilities to secure funding and profit from investments in undergrounding power lines. As much as this might cost, the loss of economic activity in states due to power outage, coupled with the constant replacement of downed grid equipment, means that undergrounding is becoming an increasingly popular choice in coastal areas.

Transmission Upgrades

There are some threats to the grid that could knock out power generation resources. In these instances, having transmission lines that can bring power from other regions becomes critical. We saw this failure in the California energy crisis of 2000–2001 where market price caps limited the supply to the state as well as in Texas in 2021 where severe cold weather knocked out generating capacity all over that state. Ensuring sufficient regional electricity transmission capacity and removing market barriers to access that electricity creates a buffer for local supply challenges.

The 2018 SEAMS study by the National Renewable Energy Laboratory demonstrated the benefits of increasing the connections between the high-voltage transmission systems of the Western Interconnection, the Eastern Interconnection, and the Electric Reliability Council of Texas. The paper showed that increased connectivity could allow the three systems to better leverage wind, solar, and natural gas power sources, resulting in a benefit-to-cost ratios that reach as high as 2.9. This result, realized through sharing generation resources and flexibility across regions, points to a significant value in increasing the transmission capacity between the interconnections under the cases considered.[20]

Cybersecurity

Cybersecurity is quickly earning parity with extreme weather as the biggest threat facing distribution grid operators. There are daily attacks on energy companies with actors like Russia and China having already proven their ability to penetrate energy networks and even shut them down, as Russia did in Ukraine in 2016, and attempted to do again during the 2022 invasion.[21]

While electric companies are clearly working hard to meet this threat, it seems that there is much left to be done. A 2019 Government Accountability Office review of cybersecurity risks facing the electric grid identified a number of vulnerabilities. These included difficulties in hiring qualified security experts, limited sharing of classified threat information between the public and private sectors, resource constraints, reliance on other critical infrastructure that could be vulnerable to cyber-attack, and uncertainty about how to implement cybersecurity standards and guidance.[22]

Although employee training to identify and report phishing attempts is important, that vigilance alone is not enough. "No amount of employee training will eliminate the occasional mistake," said Tobias Whitney, Vice President of Energy Security Solutions at Fortress Information Security. "If you're relying purely on training, that's going to be woefully inadequate."[23] To be more comprehensive, electric company cybersecurity programs need to include the following:

- systematic and prompt patching for existing systems;
- use of outside security firms to evaluate, test, and advise on system defenses;
- transition to cloud-based assets to leverage of the cybersecurity expertise of third parties;
- regular cybersecurity briefings for senior leadership to ensure understanding of current vulnerabilities;
- inclusion of cybersecurity considerations early in the adoption of any new technology or system upgrade (particularly those that include industrial control systems for remote monitoring).

Grid Segmentation

When parts of the grid go down on a linear electricity line such as a lateral, every customer past that point is out and will stay that way until the problem is resolved. This hub-and-spoke design for utility distribution is the norm because it was the cheapest model as companies built out the grid. Today, electric companies are finding ways to connect those spokes to create an interconnected distribution system.

Coupled with sensors, reclosers, and smart switchgear, segmentation allows the grid to automatically detect outages and instantly reconfigure itself to provide power to the greatest number of customers at all times. A topology that features more loops and fewer radials creates a self-stabilizing system that improves both reliability and resilience. Figure 5.2 shows the difference between a hub-and-spoke approach on the left and a more interconnected grid on the right. The connections shown allow for most customers on this grid to receive power from a second source should an outage on their lateral cut them off from their primary circuit.

Electric companies should continually examine their grid for new opportunities for segmentation. Any time major work is completed on lines and substations, utilities should evaluate whether additional grid segmentation would be a cost-effective addition to that work. There is significant industry research to support the cost/benefit analysis of segmentation and the right engineering approaches for different situations.

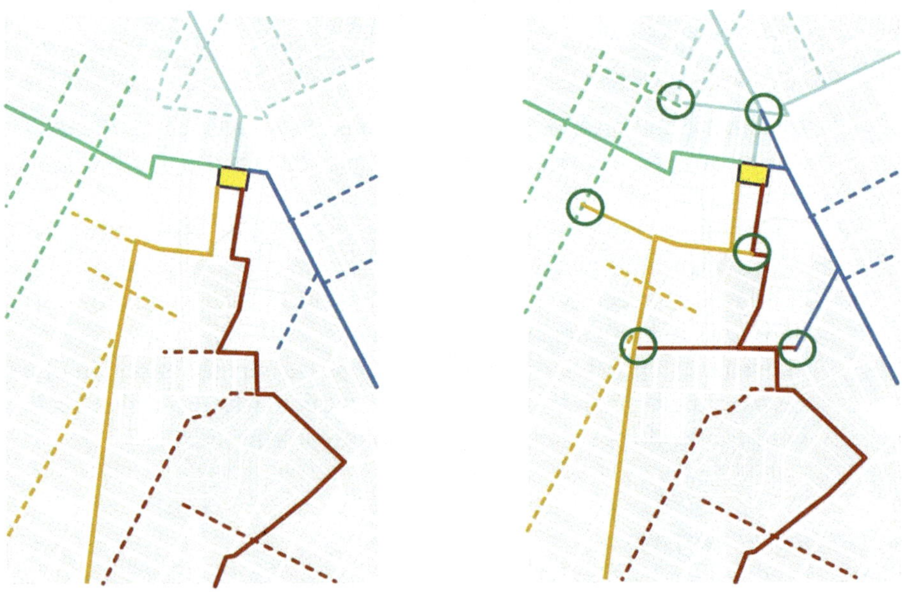

S&C Electric Company, 2021

Figure 5.2 "Radial vs. Networked Grid Topology." S&C Electric Company, 2021.

Distribution Automation and Smart Grid Technologies

Today's grid is a patchwork made up of incremental improvements to a grid that was designed in the 1890s and robustly built out in the mid-20th century. Now, the availability of sensors, digital smart meters, automated switchgear, and grid control software has given utilities new capabilities to manage their infrastructure, including the ability to:

- monitor what is happening on the grid at every scale, from generation to line operations to individual loads;
- use data analytics to predict shifts in generation and loads at multiple time scales;
- instantly respond to outage events through reconnection and grid reconfiguration;
- pinpoint areas of damage to the grid so that trucks can roll directly to the problem;
- optimize the economics of energy sourcing minute-by-minute.

Retrofitting an old system with new technology to enable the capabilities listed above can be challenging. Although smart retrofits and upgrades require large investments, the benefits of these investments have been proven time and time again – not just in lowering operating costs but also in reducing outage times and thereby reducing the economic impact of outages. Energy companies can work with their solution providers to calculate these economic benefits and make the case to regulators (and the public) that such investments will redound to everyone's benefit.

Demand Response Programs

When local distribution grid congestion or even statewide power supply becomes a problem, one of the most cost-effective ways to meet this challenge is simply by adjusting demand. The challenge, of course, is that electric companies cannot simply lower demand remotely. For this reason, these companies engage in demand response programs to provide market signals for customers to reduce energy usage when the grid is stressed.

The Federal Energy Regulatory Commission defines demand response as "Changes in electric usage by end-use customers from their normal consumption patterns in response to changes in the price of electricity over time, or to incentive payments designed to induce lower electricity use at times of high wholesale market prices or when system reliability is jeopardized."[24]

"Real-time pricing" sends price signals to try to get businesses and (to a lesser extent) residential users to shift high-energy activities away from peak hours. This is often referred to as "time-of-use rates" or "economic demand response." Other programs involve end users agreeing to a certain amount of demand curtailment if and when the utility asks (common for businesses) or allowing the utility to remote access and change thermostats (a common residential approach) – in exchange for a reduced rate or event payment. These approaches are typically called "emergency demand response."

Demand response can also be a way to increase the operational life of substations approaching their capacity limit and, therefore, deferring investment. Con Edison demonstrated this model in a 2013 rate case as part of the Brooklyn-Queens Demand Management Program.[25] Regulators have been broadly supportive of demand response programs – sometimes even requiring its inclusion in the portfolio of energy companies. To design effective demand response programs, utilities need to understand their current peak demand needs, and how these needs may evolve in the future.

Distributed Energy Resources

The transition to more clean energy in the utility system can be viewed in two ways: either as a solution to resilience challenges or as a major threat to grid stability.

Although it is true that distributed energy resources (DERs) like wind and solar have an intermittent generation profile that adds complication to supply forecasting, and these resources have resilience benefits as well. The local nature of these resources means that, paired with energy storage and the right interconnection approach, they can provide local energy when the grid at large is down. These systems also provide cost savings, carbon reduction, and improved ability to limit peak demand. Even non-renewable DERs such as energy storage and natural-gas-powered combined heat and power can support the system at large and improve overall grid efficiency by avoiding line losses.

Electric companies evaluating how to incorporate DERs should, as with demand response, understand where their congestion and peak demand challenges are coming from. They then need to partner with customers and local governments to find solutions that will meet the needs of both the utility and the site. These solutions, while expensive, are often highly flexible, allowing them to meet a broad variety of needs and use cases.

Grid Energy Storage

One of the primary challenges of grid-connected DERs is the fact that, under current IEEE standards, they must power off during a grid outage. This protects line workers from electrocution from distributed power feeding into downed lines. The exception to this is designed to be islandable, allowing it to operate fully disconnected to the grid. This works well for applications like methane-fed CHP, but for intermittent renewables, you need energy storage to shift power supply to times when power is actually needed. Systems like this are proliferating quickly to advance energy resilience at the local level and are gaining increasing consideration at the grid-level as well.

Larger scale energy storage has the ability to dramatically increase the amount of intermittent generation incorporated into the grid. It can also help balance energy supply across regions and even provide support for things like power quality, voltage support, and reactive power. While still expensive, the dropping cost curves and increasing need for grid energy storage will make it worthy of serious consideration for energy company resource planners. As the electrification of transportation soars, there may also be opportunities (like with demand response programs) to utilize the batteries in cars and trucks across the nation to help optimize the constantly shifting supply/demand equation of the grid.

Microgrids

Microgrids are coordinated groups of distributed energy resources and loads that can "island" from the grid during an outage or operate completely independent of the grid. While microgrids offer a number of benefits, their primary value is in resilience – keeping local critical loads up and running for extended periods during grid outages. These systems can also bake in the advantages of other items on this list like DERs, energy storage, and demand response and can even provide black start support for local generation resources.

While most early microgrids have been sited on the customer side of the meter, there are clear opportunities for utility-owned microgrids to support the resilience of critical substations and feeders. Ameren Illinois, a regulated electricity-delivery company, is currently operating a 1-MW substation microgrid operating at 12 kV. The system leverages wind, solar, natural gas generators, and energy storage to meet 16 different use cases.[26] Other utilities are leveraging microgrids as well, such as Pacific Gas and Electric, which is building microgrids for smaller communities at the end of long laterals.

"For us – resilience is everything to do with wildfire risk," said Bennet Chabot, Product Manager for Grid Innovation at PG&E. "Can we beat the cost of a wire upgrade with a distributed solution? This also buys down the risk of long laterals. Remote grid is finding the last house with a long enough wire where a microgrid works better than the long wire. This means fewer long wires and those wires are what cause the fires."[27]

The biggest challenge for utilities in the microgrid space so far has been in achieving regulatory approval and rate case inclusion. Since microgrids include a wide array of technologies, approval for their inclusion on the grid (or even behind the meter) can be complicated.[28] Most states have not issued specific microgrid regulatory guidance that clarifies where these systems can be sited and who can own them. States like California and Hawaii have issued clear guidance on how microgrids (on both sides of the meter) should be handled and microgrid adoption in those states is proceeding more quickly.

Line Crew Logistics

Even with all the above solutions in place, electric companies still need to roll truck and repair the grid following a storm or other disruption. Doing this quickly often means bringing in line crews from out of state to help. The first challenge of this effort is making sure that those qualified crews are available. When major disruptions occur, they often happen across a wide region – increasing competition for crews in the area.

Understanding your grid and its vulnerabilities will help you forecast how much help you really need and when. Timing is important. You have to commit to asking for crews *before* the storm both to ensure they are on-site when needed but also to secure workers before they are snapped up by other utilities.

Securing crews is one thing – having them safely housed and fed close to the area of damage is another. Energy companies need a robust logistical plan for the accommodation of and communication between line crews – sometimes from multiple states. Brian Coughlan, President of Utility Management Services, laid out some of the concerns involved, "One of the biggest challenges was what do you do with the 4,000 linesmen who have just showed up to stay and eat in your location that has no power and no spare hotel rooms. Communication is also a problem – the cell phone and radio towers are out. How do you manage the logistics of housing, feeding, coordinating and leading hundreds or thousands of out-of-town workers? The logistics of this are a major lift. You have to decide how to prioritize where they will go to work as well – which lines to address first and with how many people. You have to find a way to quantify how much each section is decimated and allocate the crews on hand."[29]

Getting this right is not just about getting the power back on quickly but also protecting the lives of the teams in the field. To prepare for this, electric companies should:

- develop multiple scenario-based plans;
- include external services like hotels and restaurants in your planning to account for their capacities and capabilities;
- communicate plans with local authorities and first responders so that there is coordination of post-event efforts;
- conduct disaster response drills, ideally including crews from outside of the service territory.

Incorporating Resilience into the Rate Base

As we have discussed, there are many options for electric companies that are committed to improving energy resilience. Perhaps the biggest challenge is securing funding to employ those measures, which is determined by how rate cases themselves are handled. Most electric companies currently only make money by investing capital. They do not make profit on operations spending, which is where a lot of the solutions above would be addressed. For this reason, many initiatives for improving resilience (in most of the United States right now) would be a money-losing endeavor for energy companies. Regulators in some states are mandating specific investments or working with energy companies on resilience programs that have a strong capital spending focus.

One solution to this dynamic is a shift from traditional cost-of-service regulation to "performance-based regulation" where utilities can profit not just from capital investments but also based on their ability to deliver results in various performance areas. This approach, which has been quite successful in Great Britain, has now been adopted in several states – including New York and Hawaii. How this shift will translate to improved resilience remains to be seen. Performance-based regulation often focuses on relatively simply metrics that can be tied directly to financial incentives – an approach that may not be practical for resilience. If done so deliberately, resilience can be worked into a balanced scorecard in a performance-based regulatory model.

A lack of familiarity with energy resilience among regulators can also be a challenge. According to Adrienne Lotto, Vice President and Chief Risk and Resilience Officer at New York Power Authority, "There is a huge gap between knowledge base and capability from the federal level down to the states. You can meet with state regulators and they are unaware of very real challenges."[9]

Of course, limitations in the knowledge base of regulators are understandable when you remember that many regulatory commissioners are in their role for short periods and, in some states, cover regulatory issues not just for electricity but for gas, water, and, sometimes, other areas as well. The focus among electric companies on resilience is relatively new, and the industry lacks clear and effective metrics for resilience performance. A report prepared for the National Association of Regulatory Utility Commissioners does an excellent job of presenting how resilience has been treated by regulators in the past and how they can best value it in the future.[30]

Electric companies looking to overcome these obstacles should consider some of the following steps:

- Carefully consider state and federal policy goals as those will determine what can be rate-based and what additional program funding can be provided.
- Partner with state regulators to explore performance-based regulation with a specific focus on resilience as one of the early performance factors.
- Use in-house or third-party experts to build out the ability to perform cost–benefit analysis around resilience investments – particularly those that would naturally live in the operational side of the budget.
- Consider developing a specific resilience strategy document and/or an annual resilience report. This will increase the visibility of the company's focus on resilience and draw regulatory attention to this as a performance area.

Utility Resilience Program Maturity Model

The following maturity model presents increasingly ambitious activities that utilities could use to deploy a more robust energy resilience program across several areas. These maturity models can be a tool for evaluating your program's place on the spectrum of energy resilience or for setting targets in future efforts. When reviewing this model, you can assume that any proactive steps in lower tier programs are included in the higher tier programs but have not been explicitly included there simply to avoid repetition.

	No program	**Reactive program**	**Managed program**	**Proactive program**	**Best-in-class program**
Stakeholders	Zero communication with outside groups	Separately track any feedback from stakeholders around resilience	Ask stakeholders about resilience specifically	Identify/include new stakeholders to better understand resilience needs	Engage stakeholders in scenario planning and simulations of resilience response
Supply protection	Leave current systems in place	Identify weaknesses during outage events and plan for system backup	Work with generators to install mitigation measures for likely threats	Diversify supply and couple mitigation measures with black start capabilities	Develop a network of grid-side microgrids to serve critical feeders during an outage event
Line protection	Roll trucks to fix equipment when it breaks	Install reclosers on laterals and implement robust vegetative management	Implement a grid segmentation plan coupled with preventative maintenance	Couple segmentation and reclosers with concrete poles and some targeted undergrounding of lines	Use big data analytics to anticipate failures and underground vulnerable portions of the grid
Cybersecurity	Rely on existing company firewalls	Identify and patch any vulnerabilities identified by attacks on the system, and deploy an employee cyber training	Look for potential system vulnerabilities through regular assessment and make cybersecurity part of the responsibility of top line managers	Implement multiple layers of security (digital/physical/human) and coordinate threat identification with other utilities	Operate isolated control systems that can connect to and control the grid if primary systems become compromised

Continued

	No program	Reactive program	Managed program	Proactive program	Best-in-class program
Response and recovery	Determine response needs after event	Evaluate response experiences to identify opportunities for improvement	Engage customers and government entities to understand restoration priorities and include that in response planning	Develop scenario plans for different threats that include template communications and plans for lineperson logistics	Coordinated wargames of response plans that include non-utility stakeholders, e.g., government and first responders
Metrics	Track SAIDI and SAIFI	Track progress toward resilience plan milestones – include in internal reporting	Develop a quantitative grid resilience metric that is approved by senior leaders	Disclose resilience metric performance to the regulator; include metrics in mid-level management performance reviews	Disclose resilience metric performance to customers and on the web site; include metrics included in top-level management performance reviews

Conclusion

As pressures mount on an increasingly fragile electricity grid, it is critical that utilities focus on resilience. Shifts in grid connected technology will bring both new capabilities and new complications. The good news is that electric companies from all over the world are developing innovative new approaches in the areas discussed in this chapter. If electric companies can actively partner with their regulators, customers, third-party experts, and each other, there are many opportunities to better weather the storms ahead.

6

Energy Resilience at the Local Level

The Importance of Local Energy Resilience

No matter how remote the root cause is of a power outage, its effects are inevitably local. The fact that a tree branch brushed against a power line on a distant hillside is of little interest to the hospital in the midst of figuring out how long it can survive on backup generation. The details of how unexpected cold can cause methane pipelines to freeze is not on top of mind for the homeowner who suddenly finds that she is without not just power but, for some reason, water as well. As we discuss throughout the book, effective energy resilience planning must be a cooperative effort between local consumers, grid operators, and government actors. However, when the efforts of the latter two actors lag or come up short, it is up to local businesses and institutions to protect their operations and the safety of the people who rely on them.

Even as energy resilience has become a popular paradigm for approaching energy security, most plans and policy prescriptions proposed have been geared toward regional security – focusing on grid-level upgrades and state or multi-state preparedness. For several reasons, we believe it often more pressing, and indeed more fruitful, to approach energy resilience on a much smaller scale: at the level of a single site occupied by a single community or institution. A shock to the energy system affects a local institution differently than it does a whole region. Although utilities and regulators are responsible for and, to varying degrees, accountable to the customers they serve, the immediate responsibility and accountability is borne by local institutions.

The National Association of Regulatory Utility Commissioners (NARUC) noted in a 2022 report that traditional investments in energy infrastructure have been insufficient to meet the public safety needs of customers at a local level:

> "The United States depends on the delivery of reliable, affordable, clean, and safe electricity. Electric utilities invest billions of dollars each year in generation, transmission, and distribution assets to meet this need. However, experiences with recent natural disasters of increasing frequency and duration demonstrate the shortcomings of this approach in the face of modern threats. Further, as customers rely on electricity for a broader range of important needs, such as transportation, as well as critical life-saving services and mission critical facilities such as water treatment, medical care, shelters, telecommunications, and more, the need to minimize the likelihood and impacts of outages grows. Against this backdrop, resilience has emerged as a key consideration to guide electricity spending, whether from utilities, customers, or taxpayers."[1]

This report, co-authored by the National Association of State Energy Officials (NASEO), is focused on the resilience value of microgrid projects and the importance of addressing energy resilience on the scale of an individual community or institution. Local consumers are on the front lines of each of the threats discussed in Chapter 1 and must manage their risk accordingly.

Finally, we know that local institutions have natural advantages over statewide or regional players, and the development of new technology and best practices make local energy resilience not just possible but cost-effective.

In this book, we define the local actor as a single community or institution occupying a defined space and able to exercise managerial control over the infrastructure in that space. This definition includes the government of a town or small city or the administration of a local institution like a hospital, university, or business. We will discuss two special cases at the end of the chapter: big cities (population of more than 250 thousand), and businesses that operate a portfolio of buildings that are spread across wide geographical area.

Energy Resilience in our Four Community Types

To illustrate the advantages that local communities and institutions have in pursuing energy resilience, we have organized the discussion in this book around four types of communities: military bases, healthcare campuses, educational campuses, and residential communities. These community types share four key characteristics that give them an advantage in achieving energy resilience, compared to a single building or a larger community:

1. **Place**: Each of these communities is defined by a contiguous footprint that allows for the coordination of energy resources for multiple loads without long-range transmission lines.
2. **Permanence**: These communities cannot simply relocate. For various reasons, where they are now is where they will likely be in 50 years. This makes long-term facility and infrastructure planning feasible.
3. **Control**: The governing bodies of these communities have control over all or part of the infrastructure and processes that drive their energy usage. This control allows for long-term energy planning.
4. **Mission**: All of these communities have facilities that play a role that is central to the mission of the organization. Whether it is a fire station, a student dormitory, an operating room, or a satellite control station, continuous energy supply is essential to carrying out the vital mission of each community.

Place: Control Over Land and Local Infrastructure

First, the land on which a community sits can offer a built-in advantage for the development of renewable energy capacity. Access to land has been found to be one of the most important barriers to the expansion of renewable energy production.[2] The development of nearly any kind of local power generation will require available space that meets certain usage regulations. Combined heat and power (CHP) plants must conform to emissions rules, wind turbines need to meet height, noise, and wildlife protection restrictions, and even photovoltaics and solar water heaters face space-use challenges based on safety and esthetics. Among the few locations that may meet all these requirements, some may still face a high degree of competition for land with agricultural, recreational, or development interests.[3] Institutions that can develop energy infrastructure on wholly owned land can avoid this competitive pressure and some regulatory hurdles. The concept of *place*, as we use it, also describes the local setting itself. The fact that the institutions we examine occupy a single, relatively small geographical space gives them built-in energy resilience advantages, especially related to on-site energy generation. When energy is generated at the same site as the institution that consumes it, the infrastructural vulnerabilities that were described in Chapter 1 are significantly reduced. Since this generation is under the control of the community itself, it can be carefully matched to the specific loads it is intended to meet. Personnel can carefully maintain the local energy system to ensure reliability and create emergency plans to cover the entire energy lifecycle, from generation to demand. The local setting allows for the kind of control that is necessary to achieve true energy resilience, as described below.

Permanence: Opportunity for Long-Term Planning

Staying put has its advantages, and the space that each of these communities occupies tends to be its permanent home. A domestic military base or hospital may eventually close, but building these institutions takes such a large investment that they are usually intended to stay open indefinitely or at least for several decades. Obviously, the closing and abandonment of a university or municipality is even less common. At colleges and universities, building assets are typically owned by the institution and managed for the same general purpose for many decades.[4] Most communities, including small towns, plan to operate at least 50 years into the future.[3] This permanence allows for long-term planning and helps justify infrastructural investments, such as energy generation equipment or energy efficiency upgrades. In all of our four community types, it often makes operational and financial sense to invest in upgrades that will return an energy resilience benefit in the short term, even if that financial benefit is not realized until the mid- or long-term planning horizon.

Control: Centralized Oversight

Power over power is another key advantage. Central governance, or control over the energy processes within a community, is a common attribute of our four community types. Decision makers must have sufficient authority over energy purchase and generation, building infrastructure, and end-use practices to engage in effective energy resilience planning. As noted in Chapter 3, this does not mean that all decision-making power has to reside with one individual or within a small group. Most likely, successful energy resilience projects will result from the coordination of multiple departments (maintenance, finance, administration, etc.) and groups of stakeholders. What is important is that the institution has the authority to lead the team and to take action on ambitious energy resilience projects. This authority is necessary to spur investments in on-site generation equipment, energy efficient technologies, continuous commissioning, and microgrids. This level of control can also make it easier for administrators to manage the demand associated with existing infrastructure. In a large municipal setting, in contrast, planners must apply regulations, zoning restrictions, or incentives to spur individual building owners to retrofit their homes or businesses for energy efficiency. In a centrally governed institution, these efforts can be mandated and coordinated. When buildings and infrastructure are owned by one organization, a community has the data and control it needs to create and implement a coordinated energy plan.

Mission: Safety and Other Essential Functions

Protecting people is at the core of energy resilience, and each of the four community types shares this priority and is focused on the safety of their constituents. Here, we use the term "constituents" to mean residents at a military base or residential community, students at an educational campus, and patients at a hospital campus. Energy resilience efforts promote safety by protecting the systems that support it, such as heating/cooling, catering, life-support equipment, and security systems. Second, it is fair to assume that cost-effective operation is a common priority among these (and perhaps all other) community types. Energy resilience measures can go far to protect institutions from financial risk. Finally, we define energy resilience as the ability of an institution to shield its core mission from shocks to the energy system. Whether educating students and conducting sensitive research, caring for the ill and injured, housing and protecting residents, or defending our national security, the mission of each of our four community types must be carried out every day of the year and cannot tolerate interruptions from an extended power outage.

The qualities listed above make each of our four community types particularly well suited for energy resilience projects. In the next section, we will briefly discuss the specific needs of each community type and the actions that are being taken to meet these needs.

Military Bases – The Energy Resilience Imperative

The importance of energy resilience on military bases is largely derived from the fourth characteristic listed above: *mission.* The operations supported by and housed within our military installations are vital to our national security in countless ways, both seen and unseen. From the protection of proprietary technologies and storage of sensitive data, to training and preparing our fighting forces, to housing and feeding their families – every activity on a military base is critical in its own way. The importance of keeping security operations unhindered is elevated in emergency situations, when a disruption in the grid may be more likely.

The Department of Defense is the largest consumer of energy in the U.S. Government, and a great deal of this energy is consumed at military bases. In 2020, military installations consumed nearly 200,000 BBtu or enough power for 3 million U.S. households over the same period.[5] If the demand for energy at these facilities goes unmet, even for a short time, the operations and the imperatives they address could be compromised. In addition to the enormous demand for electricity, energy security on domestic military bases is further imperiled by the fact that most are located in remote locations, at the end of long transmission feeders.[6]

Although we focus primarily on domestic bases in this chapter, it is important to note that the exposure of these installations to interruptions in energy supply is relatively tame compared to the risk that affects our bases abroad. The energy intensity of our forward-positioned military installations is a serious constraint to our operations.[7] The strategic options open to our military are limited by the complex logistical demands associated with delivering fuel in the field.

Military Bases – The Energy Resilience Opportunity

The characteristics of *place*, *permanence*, and *control* are at the center of the U.S. military's response to the energy resilience imperative on base. Military management has the *control* necessary to take action on this issue, and leaders undoubtedly understand the challenges posed by reliance on the commercial grid at home and far-flung distribution networks abroad. It was this understanding that led the department to start pursuing energy resilience in 2012 and to continually expand its ambition in the area ever since. Figure 6.1 lists some of the Defense Department's energy resilience programs and accomplishments over this period.

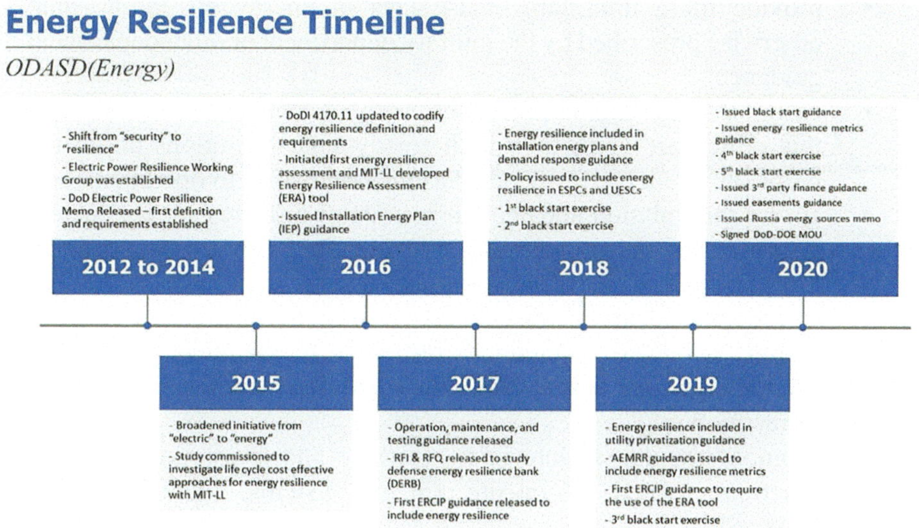

Figure 6.1 Office of the Assistant Secretary of Defense for Sustainment, Energy Resilience Program. "Energy Resilience Timeline." U.S. Department of Defense, 2019.

In 2021, the department published a memorandum titled "Metrics and Standards for Energy Resilience at Military Installations." This document lays out requirements for individual bases as well as guidance for performing energy resilience assessments. The requirements read like a list of best practices for institutional energy resilience:

- promote the use of multiple and diverse sources of energy, with an emphasis favoring energy resources originating on the installation such as modular generation;
- promote the installation of cyber-resilient microgrids to ensure the energy security and energy resilience of critical missions;
- favor the use of full-time, installed energy sources rather than emergency generation;
- by the end of fiscal year 2030, provide that 100% of the energy load required to maintain the critical missions of each DoD installation has a minimum level of availability of 99.9% per fiscal year or higher availability as this memorandum provides;

- provide for a minimum of 14 days of energy disruption, unless otherwise prescribed by the military department or other departmental guidance;
- ensure that the minimum level of energy availability standards within this policy are applied to both steady-state operations and for the purposes of energy resilience during a disruption to operations;
- ensure that mission and economic tradeoff analyses are conducted to prioritize technology solutions based on the requirements set forth in this policy;
- include black start exercises as required by law and policy;
- include adequate sustainment resources to maintain real property investments and ensure the safe operation and testing of energy infrastructure;
- promote regular maintenance, testing, and disruption prevention practices related to on-site energy systems, including backup generators;
- identify different time horizons for different phases of planning;
- base decisions on a risk analysis of threats and hazards to meet mission needs.[8]

In choosing military bases as the setting for their pursuit of energy resilience, the U.S. military can take advantage of the unique characteristics of *place* associated with these sites. This is evident from the list above, which includes requirements to promote full-time on-site generation and microgrids. Since the Department of Defense often owns a significant area of land surrounding a military base, no new acquisitions are necessary to house distributed-energy resources (DER) such as solar photovoltaics or battery storage, and no long-range transmission lines are required. At Fort Carson, solar arrays are already generating 3 MW of electricity or enough to power 800 homes on the base. In 2018, the base completed a battery energy storage facility that allows the base to reduce peak demand by an average of 9% per month. This saves the base approximately $525,000 in energy costs every year.[9]

Permanence is another common (though not universal) quality of military installations which makes them well suited for energy resilience projects. The military installations we focus on in this chapter are permanent, domestic bases, intended to operate into the foreseeable future. These bases are centrally managed by an organization capable of long-term planning so that investments can be made now toward the pursuit of future goals. As Fort Carson's Utilities Manager, Vince Guthrie, explained to *The Denver Post*,

"What this program is about is security — economic security, environmental security and national security."[10] The Department is putting money behind this mission. The Energy Resilience and Conservation Investment Program (ERCIP) began in 2008 with a $35 million budget and a remit limited to energy conservation projects. By 2021, the budget had grown to over $142.5 million, with program funding secured through 2026. Program funds now support projects related to energy resilience, availability, reliability, and economic performance.[5]

Healthcare Campuses – The Energy Resilience Imperative

As with military bases, the *mission* of hospitals and healthcare campuses is essential to understanding their need for energy resilience. Energy security can literally be a life-or-death imperative for hospitals, and energy costs often make up a significant portion of their overall budget. Healthcare campuses are among the most energy intensive facilities in the country and, combined, spend more than $8 billion on energy every year. As the DOE points out on its website, that amount of money could cover the salaries of more than 100,000 registered nurses.[11]

When significant electric loads go unmet, damage can accumulate quickly. The case of Langone Medical Center provides a stark illustration. As Hurricane Sandy settled over New York in 2012, the East River swelled and poured into the basement of the hospital. Although most of Langone's backup generators were located on high floors, the fuel tanks that fed them were in the basement. When they became submerged, they either detached from the fuel lines or were automatically shut off when their liquid sensors detected the flood.[12] The power was out and, during the next 13 hours, the hospital was forced to evacuate 300 patients. It also suffered more than $700 million in damage, much of which is attributable directly to the outage (lost revenue, interrupted research, and paying employees who were not able to work).[13]

Though it may have attracted unwanted attention due to its situation in the middle of a high-profile storm and city, Langone is hardly alone in this hardship, and versions of this story have played out many more times in the years since. In the summer of 2021, Hurricane Ida caused evacuations in several hospitals in Southern Louisiana, at a time when they were packed with COVID-19 patients. When backup power partially failed at the Thibodaux Regional Heath System there, staff members ventilated patients by hand as they were moved to another floor.[14]

The COVID-19 pandemic highlighted the role of hospitals and healthcare campuses as critical infrastructure. Already stretched to capacity by repeated waves of infections, hospitals could not afford interruption from power outages. Every hospital needs to execute an energy resilience plan to ensure that they are adequately prepared to help respond to new disasters without succumbing to them.

Healthcare Campuses – The Energy Resilience Opportunity

The Gunderson Health System, headquartered in Wisconsin, was an early leader in healthcare energy resilience. Through a decade of well planned, aggressive action, the organization has improved energy efficiency by 54% and has installed on-site energy resources sufficient to make many of their facilities energy independent.[15] In 2019, the U.S. Department of Energy's Better Buildings Program formally recognized Gunderson for accomplishments at its Sparta Clinic in Sparta, Wisconsin. The Better Buildings Program, an energy efficiency support program run by the DOE, welcomed Gunderson as an early partner. Beyond promoting energy efficiency, the Sparta site hosts on-site energy resources that Gunderson intends to manage as a microgrid, featuring geothermal heat pumps and rooftop solar panels.[16]

The success of the Gundersen case illustrates the positive characteristics of *place*, *permanence*, and *control*. Like military bases, healthcare campuses are typically permanently located on owned, contiguous space, allowing managers to develop and implement energy projects that require these characteristics of *place*. Many of these projects make use of the existing land and infrastructure of the hospitals themselves, such as the hundreds of photovoltaic panels that now cover the parking garage at Gundersen Lutheran Hospital. The need for extra land and infrastructure has also led Gundersen to seek partnerships with other community organizations to further expand its on-site generation capabilities. Gundersen Lutheran uses wind power from a facility built on land and owned by a partner, and an on-site CHP plant fed by biogas from a nearby landfill.[17]

To ensure on-site heat and power generation can meet demand efficiently, and hospitals must institute energy efficiency measures throughout the institution. The implementation of these measures relies on the strong central *control* typically seen on a healthcare campus. At the University Medical Center of Princeton in Plainsboro, New Jersey, planning before its opening in 2013 and careful management since then allowed the entire hospital and all of its operations to be powered with on-site generation. Instead of using its on-site systems to back up power from the regional grid, the hospital uses grid power as the backup. Should both systems fail simultaneously, there are battery arrays that provide a third layer of resilience for essential equipment.[18] This accomplishment is the result of a systems approach that is only possible if all components are planned in concert and carried out with central oversight and control.

Educational Campuses – The Energy Resilience Imperative

Every college or university campus is like a small city unto itself, with a wide array of functions and facilities, and all the attendant energy requirements. The importance of energy resilience on college campuses again comes down to *mission*. Residential students need HVAC, hot water, light, and electricity in their residences, and cafeterias capable of serving three hot meals a day. Colleges cannot simply close if the power goes out and, as we discussed earlier, the provision of a safe and secure environment for student residents is central to the mission of any residential campus. The mission of many universities also includes ongoing research, much of which must be powered without interruption to preserve experiment parameters.

On the same night that saw the evacuation of the NYU Langone Medical Center, several of the university's research centers also lost power, including the Smillow Research Center, which houses long-term live animal research projects on heart disease, cancer, and neuroscience. Freezers thawed and animal areas flooded, causing years of research to be lost. "It is not very different from someone losing their entire home. For scientists, their research is their lifeline," Dr. Dafni Bar-Sagi, NYU's Senior Vice President for Science and Chief Scientific Officer told a reporter. "For someone who started three to four years ago, and just got to a point to launch their research program, it's time to rewind and start from scratch."[19] Of course, severe weather and power outages can displace students as well as their research. When Hurricane Ida hit New Orleans in 2021, it left the campus of Tulane University largely without power and other critical services. The university had to make the difficult decision to evacuate 1841 students via bus and house them in hotels in Houston. Many more students were asked to evacuate and return home while the university repaired damage and sought to restore power.[20]

Educational Campuses – The Energy Resilience Opportunity

Driven by the need to control costs and a desire to promote environmental sustainability, energy efficiency is a priority for nearly every college and university in the country. Many schools are realizing that to protect their mission, they must go beyond efficiency to achieve broader energy resilience. These campuses take advantage of the characteristics of *place*, *permanence*, and *control* to find success in energy resilience. *Permanence and place* are keys to Cornell University's strategy to become energy resilient, and the university is now able to meet its entire peak load of 35 MW with on-site generation. A CHP plant generates most of the university's electricity and produces 90% of the heat used on campus. The school's cooling needs are met with a deep water lake-source cooling facility. When the power from the regional grid goes out, the university can continue to operate uninterrupted.[21] The school is also investing heavily in solar power and heating on campus and is taking advantage of its rural setting to build solar farms at research sites spread around central New York. Cornell is pursuing a goal to meet 100% of its energy needs with renewable sources by 2035 and can currently meet 100% of the load of its main campus with renewable energy on sunny afternoons.[22]

Even in urban environments, colleges are finding ways to take advantage of their centralized *control* over infrastructure to plan large-scale energy projects. Lacking the surface space enjoyed by Cornell, NYU installed a CHP plant underground. While strolling through the plaza at 251 Mercer Street, a visitor would never know that beneath their feet lays a state-of-the-art, $125 million CHP plant that took over two years to build. It provides electricity to 22 NYU buildings and both heat and hot water to a total of 37 buildings. This plant kept the lights on at NYU's Washington Square campus throughout Hurricane Sandy, when the rest of lower Manhattan was in the dark. John Bradley, NYU's Assistant Vice President of Sustainability, said at the time that "The entire neighborhood was dark -- everything. And then there was us - It really was a little surreal."[23] Without this plant, NYU would not have come through Hurricane Sandy as a success story. Although it may not seem like much compared to the capital cost, the $5–8 million in energy cost savings NYU is realizing from the plant is considerable.[24] Still, it took a central administration with a tolerance for long-term payback to make the project a reality and to realize the enormous energy resilience benefits it affords.

Residential Communities – The Energy Resilience Imperative

Of our four community types, residential communities may seem like the least likely to attain the kind of high-level energy resilience described in this chapter. However, these communities resemble campus environments in important ways. We describe a community as defined by contiguous land and the institution that occupies it. In a small town, this institution is the local government. Like the other three community types, the *mission* of a local government depends on energy resilience. Again, the safety of constituents is a primary concern. In an emergency situation, the local government is the first and most important safeguard for local residents. To perform this function, the municipality must maintain power at fire and police stations, emergency shelters, and emergency operations centers. If these facilities go dark along with the rest of the town, the government will fail in its mission when it is needed the most.

Recognizing this imperative in the wake of Hurricane Sandy, New Jersey sought to leverage federal funds to improve energy resilience at the municipal level. This spending included grants from the federal Hazard Mitigation Grant Program (HMGP) for projects to improve resilience in life safety facilities like police and fire stations, shelters, emergency operations centers, and water supply and wastewater treatment plants.[25] The grants available through the HMGP totaled $100 million, but in a demonstration of the extent of local governments' interest in energy resilience, New Jersey received over $469 million in requests for the funding.[26] The state has since developed a program of "Town Center Distributed Energy Resources (TCDR)" microgrids covering critical facilities in small municipalities. In 2021, state regulators approved $4 million for eight new TCDR microgrid designs. The designs will cover both Category IV facilities (those providing critical services such as hospitals, fire stations and water treatment facilities) and Category III facilities (those that could serve as temporary shelters during an emergency).[27]

Residential Communities – The Energy Resilience Opportunity

Residential communities have many of the same characteristics that we have used to identify energy resilience opportunities in our other community types. Municipalities have a core *mission* that depends on continuous power. They have *place* and *permanence* characteristics that are well suited to energy resilience planning. Municipalities typically possess a fair amount of publicly owned land and infrastructure on which new energy projects could be developed. Residential communities may also be the most permanent among our four types. It is not uncommon for municipalities to make management plans that extend 50 years into the future. Even with the space and permanence to support large-scale energy resilience planning, municipalities may sometimes lack the *control* necessary to carry them out. Our analysis of residential communities in this book focuses on public infrastructure – buildings and other assets that are owned by the municipality or community. The kinds of projects funded by HMGP grants in New Jersey will affect public infrastructure and will not require projects or behavior changes from private citizens. However, it should be noted that the support of residents can make a big difference in energy resilience planning at the municipal level.

The level of control wielded by publicly elected officials will seldom match that of a central hospital or university administration. Nevertheless, an organized and engaged constituency can make a town supervisor or city council a potent agent of change when it comes to energy resilience. The residents of Montgomery County, Maryland, have shown the power of public support. In 2012, a derecho blew through the county, leaving 71 county facilities and 250,000 residents without power for several days. The county had already established a goal to reduce emissions 80% below 2009 levels by 2050, and in the wake of the storm, the community started looking for projects that would help them not just to accomplish their sustainability goals but to build a more resilient energy system.[28] In 2018, the county completed work on two microgrids, one covering their public safety headquarters in Gaithersburg, and a second at a nearby jail. Both were created using an "Energy as a Service" model that allowed the county to build the projects without any capital expenditure. The county is paying for part of the projects through the energy savings it creates. The project developers that funded the project will be repaid through sale of the energy it generates on the wholesale markets, renewable energy credits, and potentially other ancillary services such as demand response.[27]

As an aside, the county microgrid projects were the first to receive PEER Platinum certification. A third-party accreditation, the Performance Excellence in Electricity Renewal (PEER) program, is designed to recognize high performance, resilient power systems, the way the LEED certification recognizes performance in building sustainability. If this program can gain the kind of recognition and cache the LEED program has attained, it may create an added incentive for institutional investment in energy resilience.

As of this writing, Montgomery County has begun developing another microgrid, this time as part of the new "Brookville Smart Energy Bus Depot." The county is transitioning 44 of its busses from diesel to electric, and the microgrid project will allow them to recharge at the depot using 100% clean energy.[29] Transportation is another essential service in emergency situations, and the bus depot microgrid will make this service much more resilient to energy disruptions, whether from grid blackouts or fuel price volatility.

Local Resilience Special Cases: Cities

The services that protect a large city (for our purposes – a city with a population of more than 250 thousand) are different enough from those that serve a small municipality that they require distinct energy resilience strategies and do not fit easily into our model of local energy resilience as defined above. As we saw in the New Jersey and Montgomery County cases, the layouts of smaller municipalities often lend themselves particularly well to microgrid projects. For example, if a town hall, fire station, and police station are all located in reasonable proximity to each other, one microgrid project could protect all three during a grid outage. This hypothetical town may decide to construct one additional microgrid to serve a hospital or water treatment facility located away from the town center.

A big urban area, on the other hand, is a network of infrastructure designed to serve a large and densely populated area. Building a microgrid for one police station does little for the nearby neighborhood that is not served by the officers there. Cities also serve a mission that extends beyond their own boarders. For example, most big cities host international transportation and shipping hubs in the form of airports and seaports. A big metro area requires a thorough, multi-faceted energy resilience plan that combines facility-level projects with other measures like distribution grid improvements, emergency services planning, and even upgrades to non-energy infrastructure like floodwalls. In this section, we use recent actions in New York City and Houston (two cities that have repeatedly contended with extreme weather) to discuss a few of the energy resilience tactics that are specific to big cities.

Hardening of Non-Energy Infrastructure

Big cities may need to focus on hardening their non-energy infrastructure to protect the resilience of their energy systems. In the nearly 10 years that have passed since Hurricane Sandy, New York City has done extensive resilience planning and preparation. The city has focused much of its attention on the factor that caused the most damage during the hurricane – coastal flooding. As the case of Langone Medical Center and its basement fuel tanks illustrated, flooding can have devastating effects on energy systems. On the other side of the same coin, power outages can leave those displaced by flooding without the essential services they need to stay safe. The city published zoning plans to reduce flood risk in vulnerable areas, partnered with the Army Corps of Engineers to harden coastal infrastructure in the Rockaways, and fortified protections for the waterfront in Manhattan with the Lower Manhattan Coastal Resiliency Projects.

As it has repeatedly over the past two decades, climate change upped the ante in 2021, as Hurricane Ida delivered more rain in 1 hour than the city had ever seen (a record that had been set only two weeks prior to Hurricane Henri).[30] Water poured into New York subways and basement apartments, and more than 150,000 homes in the tri-state area were left without power.[31] New York City has attacked resilience from many angles since Sandy, but it is clear that there is more to do. Being so far up the coast, New Yorkers were not used to thinking of themselves as in the line of fire for hurricane damage before 2012, but now the biggest city in the U.S. has no choice but to face every hurricane season as if it could bring another Sandy or Ida.

The recent experience of Houston also illustrates the importance of investing in non-energy infrastructure to ensure energy resilience. When Hurricane Harvey struck Texas in 2017, it left more than 330,000 customers without power. Most outages were caused directly by rain and flooding: contaminating fuel at power plants, damaging transmission infrastructure, and preventing power plant staff from reaching their workplaces.[32] The formula that made Houston's rapid growth possible – low taxes, low regulation, and relatively free reign for developers – had put many of the city's neighborhoods at the risk of flooding. As Susan Carroll of the Houston Chronicle wrote once the flood waters had receded:

> "Houston's deference to developers was evidenced by the thousands of homes built in known flood plains and floodways, clogging the path of rushing floodwater and causing it to rise. Developers not only had built out to the bases of the reservoirs that once sat on the far western flank of Harris County, but inside them — within their flood pools — and right up to their emergency spillways. Over time, Houstonians became desensitized to the risks of living about 50 feet above sea level."[33]

The National Weather Service estimates that Harvey caused $125 billion worth of damage, on par with the cost incurred by Hurricane Katrina. Over 300,000 structures were flooded, and 40,000 flood victims were forced to evacuate their homes and seek refuge in shelters. The storm was directly responsible for at least 68 deaths in the United States, over half of which were in the Houston metro area.[34]

The city of Houston has a long list of infrastructure projects it would like to undertake to mitigate the effects of future flooding, including the creation of a new reservoir, large-scale detention basins, and new "Coastal Spine Storm Surge Protection System" that would protect Galveston Bay from storm surge.[35] Each of these projects is essential to energy resilience in a city that has suffered through such extensive outages from flooding.

Marissa Aho, the Policy Director and Chief Resilience Officer for the City of Houston Texas, explained it to us this way: "Effective resilience planning is about changing the way we do things – fundamentally – to be proactive instead of reactive. We use the word "precovery" instead of recovery. We know where the vulnerabilities are, but the systems are set up to address things when they go wrong – not to prevent them from going wrong in the first place. Spending the money ahead of time instead of after creates huge savings."[36]

Microgrids and Distribution Grid Improvements

Although a single microgrid will not yield the same kind of benefit for a big city that it can for a smaller municipality, the strategic deployment of microgrids is an important part of urban resilience planning. Instead of covering a single cluster of municipal facilities in a city center, microgrids can be developed at strategic locations throughout a city to protect vital services and the local economy. In the period since Hurricane Sandy, the New York Economic Development Corporation developed 11 small microgrid projects for small businesses throughout the five boroughs as part of its Resilience Innovation for a Stronger Economy (RISE) program.[37] As of this writing, there is a microgrid project underway at Hunts Point, which is home to the largest of New York City's six major food distribution centers. The distribution center sits on the coast in the increasingly vulnerable 100-year flood plain – a significant risk to the food supply for the city's eight million residents.[38] New York's JFK airport is working on a new, state-of-the-art Terminal One and intends for it to include a microgrid that will reduce energy use by 30% and contribute to the airport's goal of achieving 100% renewable energy by 2030.[39] Going forward, the city should consider working to cover more critical city facilities with hardened microgrids. They could also work with the state PUC and with Con Edison to upgrade energy distribution infrastructure to be ready for winds, rains, and storm surge that previously would have been considered too rare to justify the investment.

In partnership with the utility CenterPoint Energy, the city of Houston is embarking on an "all-of-the-above" approach to energy resilience, which includes upgrades to the distribution grid. In 2021, Houston faced not only the destructive winter cold discussed in Chapter 1 but also Hurricane Nicholas in the fall, a storm that left 460,000 people without power in Houston alone.[40] In response to these disasters, and with the knowledge that there are likely more on the way, the City of Houston and the CenterPoint teamed up to launch a program they call "Resilient Now." They described the program in a press release as a "first-of-its kind collaborative framework to develop a regional master energy plan to enhance local power resilience across the greater Houston area."[41] The program is designed to build on recent actions aimed at energy resilience.

In 2021, the Texas Legislature authorized T&D utilities (like CenterPoint) to engage in various activities to prepare for severe weather events, including demand response and procurement of mobile backup generators for emergency deployment to customer sites. The Resilient Now program will give the city and utility a chance to engage as partners in a thorough resilience planning process and to identify measures to be taken over the next 10 years to prevent the kinds of impacts that recent storms have had on the area. Houston has ambitions to include sustainability in the eventual energy master plan, and leaders are looking into opportunities to build out EV infrastructure, advance residential weatherization, and reuse brownfields for new renewable generation. Houston Mayer Sylvester Turner put it this way when announcing the program: "My administration set out to make resilience a foundational priority because, as the past six years have shown us, we continue to face increasingly challenging weather events of an unprecedented magnitude. We must all work together to address power vulnerability and insecurity across every community in our area."[36]

It remains to be seen what shape the final energy master plan will take in Houston, but the wide array of projects being debated bode well. City resilience plans need to match the complexity, geography, sociology, and unique risk profile of the city itself. Importantly, Houston is already following a path that aligns with the recommendations presented in this book. First, the city is initiating a deliberate master planning process focused on energy resilience, as described in Chapter 2. Next, they are addressing all three layers of energy resilience planning: local (Houston itself), grid (CenterPoint), and government (the Texas legislature).

Marissa Aho explains her approach this way: "I look at resilience by scale. If you do not improve resilience at each scale, none of the scales are resilient. People need to be resilient. Private businesses need to be resilient. Governing bodies need to be resilient too, but we can't be bound by our municipal boundaries, because they don't matter to nature. That's where regional partnerships are important. You need to pursue to goal of resilience at every scale, as a coordinated effort."[36]

Local Resilience Special Cases: Portfolios of Distant Facilities

Energy resilience is a unique challenge for businesses and other institutions operating a portfolio of facilities across a wide region, like a chain of retail stores or a network of warehouses. The risk profile will be different in New Mexico than it is in New Hampshire, and a Public Safety Power Shutoff in California may be hard to manage if a company's attention is on an approaching storm in Florida. Retail stores may rent their space, meaning they lack the *permanence and control* quality that we identified above as an advantage for energy resilience. Whether retail chains are associated with the kind of *mission* that necessitates a high level of energy resilience may depend on the type of store. Business continuity is always a worthy goal, but a clothing store does not necessarily need to stay open through a hurricane, a time when few people are likely to be shopping anyway. A grocery store, on the other hand, not only provides an essential service to the community but also is particularly vulnerable to power outage. Without backup generation that can cover its refrigeration load, a store can lose a huge portion of its inventory to spoilage in a matter of hours.

Stop & Shop, a major Northeast Grocery Store Chain, took the initiative to advance its energy resilience, installing microgrids at 40 of its stores across Massachusetts and New York. The microgrids will be fed by fuel cells from Bloom Energy and are designed to run 24 × 7. Under normal operation, the fuel cells will provide cleaner, less water intensive energy to the stores, and when the grid is out, the systems will island and cover critical loads in the store, including freezers and refrigerators for food and prescription drugs. Stop & Shop piloted this model with Bloom at two stores in New York starting in 2014, and having proven its effectiveness, the chain is ready to expand the concept in more of its stores.[42]

This kind of repeatable microgrid deployment shows potential not just for improving resilience but also as a profitable business model. New ventures are starting to sprout up with this model as their sole focus. GreenStruxure, a joint venture between Schneider Electric and Huck Capital, announced in 2021 an infusion of $500 million from Blackstone to finance modular green microgrids.[43] The modular, standardized approach of GreenStruxure's offer is well suited to installation across a portfolio of facilities since the solution can be engineered once and needs little tailoring from site to site. In another sign of the market's confidence in this type of solution, Goldman Sachs in 2022 announced a $75 million investment in GridPoint, whose software platform allows companies to manage their energy use across a portfolio of facilities. Until now, GridPoint has been focused on HVAC and lighting controls, but it plans to use this new influx of capital to expand its offer to support EV charging, on-site energy storage, and other solutions.[44]

The ability to actively manage the flow of energy across a portfolio of facilities is at the heart of how another industry is thinking about energy resilience – data centers. In the data center industry, this kind of easily repeatable design is almost as important as reliability and resilience. Data centers are the warehouses of networked computing equipment on which nearly every technological task relies, from video games to banking to managing the energy grid itself. The industry has seen enormous efficiency gains over the past decade. Even as data center capacity grew 550% from 2010 to 2018, total data center energy use around the globe has increased only 6%.[45]

Despite these advances, data centers still consume huge amounts of energy – as much as 2% of total grid power around the world – complicating the process of ensuring reliability and resilience. One thing that has not changed much in decades of data center development the main tool employed to that end – diesel generators still dot the outside of nearly every new data center. Typically, data centers pair diesel generators with uninterruptable power supply (UPS) systems powered by batteries. These systems provide near-instantaneous power to the data center when the utility feed is lost, but they are only designed to cover the load for the short period of time while diesel generators power up. UPS systems are also used for power quality correction for the diesel generators and/or in normal operation.

For a variety of reasons, the industry is looking beyond the diesel/UPS model and is examining new models to enhance their energy resilience. One of the key motivators for this evolution is sustainability. Many of the biggest names in the data center industry are corporate giants such as Microsoft, Amazon, Apple, Google, and Meta. Diesel generators do not align with the ambitious sustainability goals these companies have set for themselves, even if the generators only run for a short time out of each year. Most of the other biggest players in the data center industry count these big five tech companies among their customers and, therefore, must adhere to the same sustainability goals to retain their business.

Most of the new resilience solutions the data center industry has deployed to improve resilience and sustainability include some form of microgrid. A microgrid supports the incorporation and prioritization of clean DER into a data center's emergency power supply system, even if they do not fully replace the diesel generators on-site.

Data center operators are also turning toward microgrids for the advanced functions they can perform to both optimize sustainability and drive down costs. These functions are possible due to the "always-on" nature of a microgrid and the DER that it comprises. For instance, a data center operator could use microgrid control software to choose to consume power from DER when the cost of grid energy is highest or to take advantage of opportunities for demand response. Data center operators are exploring this kind of optimization for carbon emissions as well. They could choose to consume energy from the grid when the production of grid renewables is high, for instance. We will get into more detail on these functions in the *Microgrids* section later in this chapter.

Tactical Approaches to Local Energy Resilience

The cases and examples in this chapter have referenced many of the tactics for local energy resilience planning, including public policy, infrastructure projects, and the application of new technologies. In this section, we will give a survey of these tactics, in two main buckets: energy efficiency upgrades and on-site energy generation. These are tools local institutions have at their disposal as they create a comprehensive energy resilience plan as described in Chapter 3.

It is important to remember that it pays to be expansive in your investment strategy, considering efficiency, generation, and emergency preparedness as interdependent. Changes in one part can dramatically affect all other parts. As you develop your plan for energy resilience, it is possible to deliver an overall value that is more than the sum of individual projects.

Energy Efficiency Approaches

Although energy efficiency is not synonymous with energy resilience, there are several reasons why energy efficiency projects may be valuable first steps in a local energy resilience strategy.

1. Lower usage = lower risk

Using less energy means less risk exposure to various kinds of energy system shocks, such as short-term outages, short-term price spikes, changes in demand charges, long-term price increases, and fuel supply interruptions. Reducing total energy-buy means energy consumption will have a smaller total footprint on your bottom line and less potential to disrupt your organization's finances in times of higher prices. At the same time, stored energy (in both stationary and automotive fuel tanks) will last longer, allowing vehicles and backup generators to operate longer when fuel supplies are interrupted.

2. Decreased need for on-site generation

Reducing total energy load means that less on-site generation and storage is needed to operate independently from the grid. On-site generation infrastructure can be expensive. Reducing energy demand minimizes the size and expense of installing and maintaining on-site generation systems.

3. Efficiency savings can help fund investments

The cost savings associated with efficiency and conservation measures can help offset the cost of investing in new energy infrastructure like on-site generation, storage, and management software. Some organizations are willing to invest to improve their environmental sustainability. Many are willing to invest to reduce their risk of power loss. *All* organizations are willing to make investments in efficiency that will yield attractive paybacks. Although the parameters on those paybacks differ, immediate financial reward from a good investment is always a motivator. As your organization weighs the value it puts on those first two priorities, you can counter real or perceived costs with highly quantifiable savings from energy-saving projects. This kind of bundling can make a power storage or solar project revenue-neutral over a certain number of years when paired with efficiency investments.

Tailoring Demand Reduction to Meet Resilience Goals

The technologies and strategies you use to improve energy efficiency should take all of your goals into account. Cost savings and environmental performance are often top priorities of energy efficiency projects. As such, projects that have the largest impact on total energy usage and the quickest payback are frequently taken on first. However, in the context of energy resilience, the "low hanging fruit" approach may not be the wisest course when choosing which projects to implement first.

In a crisis (whether an outage event or another problem that limits access to energy), certain technologies will be critical. Lighting, heating, and, in some cases, refrigeration will be top priorities. Mechanical equipment, such as pumps or communication equipment like telecommunications switching stations, may also be vital. As part of your overall resilience plan, prioritize buildings and specific end-uses for energy so that you can develop a triage system to meet the most critical loads first. The prioritization of end-uses at your institution can inform your decisions around where to perform energy audits, and which facilities and systems to target with efficiency and conservation measures.

As an example, let us look at two retrofit opportunities for a community that has some on-site generation but not enough to run all of its facilities in the case of a grid outage. A public library needs upgrades to its boiler and chiller. Both pieces of equipment are working, but they are old and inefficient. The project would have a total cost of $45,000 with a payback of 3.5 years. Meanwhile, the county could also spend that $45,000 upgrading the lighting in the police station, fire station, community center, and City Hall. These buildings already have relatively efficient T-8 lighting, but an upgrade to LED lights would save a great deal of energy and have a payback of five years. If the goal of energy management is purely about cost savings, the boiler and chiller upgrades are the better project. However, heating and cooling the library will be a lower priority during a grid outage than lighting government buildings essential to emergency response. Given this community's limited on-site power options, lowering the total energy use of that lighting is a higher priority than the energy usage at the library. They are both good projects with a solid payback, and, hopefully, both can be implemented. If one has to be chosen, the lighting upgrade will do more to improve energy resilience.

Continuous Energy Performance Improvement

Although discrete energy efficiency projects may be incorporated into an energy resilience plan, energy efficiency can and should be addressed as a continuous part of facility management. A portfolio-wide approach should be used to identify improvement opportunities, implement top priorities based on available budget, and track performance. Any organization that follows the eight basic steps listed below will achieve what we call *continuous energy performance improvement*. As equipment fails, technology improves or needs change, and your facilities will continue to perform better and yield additional benefits for energy resilience.

You will notice some overlap between this approach and the higher-level approach discussed in Chapter 3. Energy resilience planning is a broader effort than simple energy management, but several of the best practices presented below are common to both, and the work done in the energy management area would logically roll up to your larger energy resilience planning efforts.

Eight Steps to Continuous Energy Performance Improvement

1. Portfolio-wide benchmarking: Developing an energy baseline allows you to understand the starting point against which you will set goals and measure progress. You can use historical utility data and to analyze a building's energy usage and make important assumptions to inform potential projects. The most common approach to energy performance benchmarking is to enter three years of historical building utility data into performance benchmarking software. ENERGY STAR's Portfolio Manager (www.energystar.gov/benchmark) is an excellent and free online tool for energy performance benchmarking. It will provide figures for weather-normalized energy intensity per square foot for each building in your portfolio. It will also give you a 1–100 score for most space types, telling you how efficient that facility is compared to others like it across the country (again, weather normalized).

2. Targeted building energy audits: The purpose of an energy audit is to identify opportunities for energy savings based on the current energy profile at your institution. These opportunities are referred to as energy conservation measures (ECMs). Before commissioning any audit, you will want to become familiar with the ASHRAE (formerly the American Society of Heating, Refrigeration and Air Conditioning Engineers), tiered classification system for energy audits.[71] These categories include the following:

- *ASHRAE Level I Audit*: This audit involves walking through the space to identify areas of energy waste or inefficient technologies. This kind of audit does not involve actual measurements and does not result in any financial analysis of the ECMs it identifies.
- *ASHRAE Level II Audit*: A level II audit is a more involved process that evaluates energy performance opportunities in the building envelope, lighting, HVAC, domestic hot water, plug loads, compressed air, and process uses (for manufacturing, service, or processing facilities). The audit team collects data about the technologies, controls, and

management practices in place. This data is used for financial analysis of each ECM.

- *ASHRAE Level III Audit*: Level III audits, or "investment grade audits," typically look at specific ECMs identified in a lower-level audit and conduct much more in-depth analysis of the anticipated effect of that ECM. These audits are expensive and only recommended for institutions considering large investments.

3. Targeted building commissioning and re-commissioning: When a building is first brought online, facility managers put it through a commissioning process. This process involves testing each energy-using system in the building to ensure that it works as designed. By extension, retro-commissioning is simply performing a commissioning study on an existing building. This kind of study can spotlight a number of energy-draining problems, such as incorrectly installed equipment, poorly calibrated controls, incorrect overrides, and failing equipment.

4. Identification and prioritization of energy conservation measures (ECMs): Once you have completed your analysis, you will have a list of measures to consider. If you performed a Level II or higher audit, you will also have financial information around what those projects will cost and their probable payback. It is likely that you will need to select some subset of these ideas, due to budgetary constraints, and competition from other potential resilience projects related to on-site generation or emergency planning.

5. Determination of financing approach: This step should take place simultaneously with the prioritization of ECMs since the funding mechanisms used will affect what is possible to pursue. In Chapter 3, we discuss some of the main funding sources available for facility energy projects. There are additional funding sources unique to specific organization types (like institutional endowments or bond issuance) not discussed in this book. Whatever funding approach your organization takes, be prepared to have concrete estimates of anticipated energy savings (with clearly-stated, realistic assumptions) before determining your financial approach.

6. Selection of energy efficient technologies to be deployed: Once you have determined *which* ECMs you will implement and *how* you will pay for them, it is time to select specific technologies and suppliers. National labs such as Lawrence Berkeley National Laboratory, Pacific Northwest National Laboratory, and the National Renewable Energy Laboratory conduct high-quality analyses of new technologies with no bias toward or against specific product types or manufacturers. As with all major purchases, it is important to engage multiple potential vendors – regardless of whether you plan to make your purchase through a proposal process or by buying directly from a single source.

*7. **Implementation of ECMs:*** Implementation is a critical project management step – and one that can present challenges. Various contractors, manufacturers, and facilities team members must be coordinated to ensure that the right people are installing the right technologies at the right facilities at the right time. A professional project manager should oversee the process, and deadlines should be explicit in the contracts of equipment vendors and third-party installers.

*8. **Performance tracking of ECMs:*** Once implemented, you should closely track the performance of new energy technologies or procedures. Tracking is critical for two reasons. First, adding new technologies or approaches into your system can have unanticipated consequences on other systems if done incorrectly. Tracking the performance of your new systems is a way to ensure that you get what you have paid for. Second, demonstrating the cost savings from ECMs is an excellent way to build internal support, which can lead to further investment in energy management and demonstrate the value of your team to cost control.

Once the ECMs identified in your analysis phase have been addressed or consciously set aside, it is time to begin again. This should not mean performing energy audits every year. Indeed, the cycle to fund and implement new ideas is typically longer than one year. It does mean that your team should constantly be spending some of its time on one of the eight steps above for each facility. Some of them, like performance benchmarking, should be ongoing efforts.

Distributed-Energy Resources

The ability to generate and store power on-site and deliver it to where it is needed when the grid goes down is an extremely potent tool of energy resilience, and one of the key advantages that local institutions have in pursuing it. This section will discuss how to understand your on-site generation needs and describe some of the most common on-site energy technologies to provide you with context for evaluating options. Finally, this chapter will discuss microgrids – a combination of infrastructure and information technology that makes independent energy operation possible for local communities and institutions.

Evaluating Your Power Requirements

The process of measuring critical loads in your facility and evaluating the cost-effectiveness of on-site energy resources can be organized into five steps. Steps 1–3 can be done in any order, depending on what information is easiest to access, or what information is needed most urgently:

1. ***Benchmark power needs:*** Before making major investments in on-site power, you need to understand how your portfolio of assets uses energy. Understanding the loads of various buildings in your portfolio, and systems within those buildings, is essential to following four steps. You will need to understand how average and peak demands, and how demand changes through the course of a day, week, and year. The energy benchmarking process is covered under *Energy Efficiency Approaches* earlier in the chapter.

2. ***Understand constraints and limitations:*** It is important to understand all the constraints that may impact your selection of on-site DER, from technical to social. Potential technical limitations include the physical space available for new equipment, existing electrical distribution infrastructure, feasibility of establishing an interconnection with the utility grid, and available natural gas infrastructure. There may also be regulatory constraints to consider, including net metering tariffs, interconnection rules, right-of-way limitations on new distribution infrastructure, and commercial generation regulations. On the financial side, you will need to assess how the cost of delivered energy from a new DER project stacks up against the current cost electricity from the utility, and how this may affect the bankability of your project based on its funding model. For instance, if you are considering using an Energy-as-a-Service model to avoid capital costs, you will need to ensure that prices and revenue streams are acceptable to you and your project developer, respectively.

3. Choose a level of resilience: Setting a target level of energy resilience may be the most important step you take in this process. It will create parameters around your entire effort. This step could easily come first or second – depending on whether you want to let an understanding of your portfolio drive this decision, or let this decision drive your portfolio evaluation. Review the energy program maturity model in Chapter 4 for a more in-depth discussion of levels of energy resilience management. For the purposes of scoping a DER project, resilience goals can range from bare bones coverage of a single critical function to the protection of the full facility loads, which may include the export of surplus generation to the grid under normal operation.

4. Target certain facilities or functions to be sustained by on-site generation or emergency power: Once you prioritize your loads, you can make sure your investment in DER matches your need. There are a few general questions that can help you to properly prioritize the loads of the buildings at your facility and the systems within them. First, what is needed to ensure the immediate safety of the people in your organization and to continue your core mission? For a hospital, the answers to these questions may be one in the same. A university may have to consider sensitive research and animal subjects in addition to the safety of its students and staff. Next, ask yourself how much energy is needed to continue these functions during a grid outage. Use the benchmarking from Step 1 to determine the answer. Finally, determine what you need to support a speedy return to normal operation when grid power is restored. Utilities have dedicated and capable engineers and repair personnel who work long hours in dangerous conditions to restore power when it goes out. When they succeed, many organizations are still not able to quickly return to normal operations because they have not made a plan for ramping back up. After protecting human safety and organizational mission, dedicate additional power to support the operational staff that will manage and effect this transition back to normality.

5. Select the right technologies to meet demand: Once you understand your desired level of resilience, your energy load patterns, and the limitations and opportunities facing this project, you can finally ask the question, "So, what should we install?" There is a wide range of DERs available in the market, and it is important to select technologies with capabilities that closely match your needs. The next section surveys some of the major distributed-energy technologies in use today and where each of them makes the most sense. You will need to match up what you learned and decided in Steps 1–3 with technical advice on how each technology can support your system's energy resilience.

Before we describe some specific categories of DERs, let us review some of the key considerations to keep in mind when selecting them.

Interconnectivity – Your ability to interconnect multiple generation sources and loads is critical when it comes to certain energy resilience strategies. If you cannot create a local microgrid that can operate in island mode, you have significant limitations around how distributed-energy sources can be used. In that case, each generation source would need to power a discrete circuit or building and have dedicated wiring within that facility to function in case of grid disruption. If you can interconnect loads and generation sources in a microgrid, then you can produce power anywhere within that microgrid to serve any connected load. This interconnectivity refers to both the physical wiring of energy resources and loads into a network, and the digital connection among all of the above. The digital networking of all DERs and loads allows for SCADA and microgrid control software to monitor and control the system. Without this connectivity and control, your on-site energy system cannot perform functions that vastly improve its value and payback, such as actively directing power to where it is needed, and interacting with the grid during normal operation to take advantage of financial opportunities like demand response and time-of-use rates.

Scale – The amount of power you are trying to produce will significantly impact your generation choices. If small amounts of power are needed, a photovoltaic system (possibly with battery storage) may be the right answer. If you need large amounts of base load power, a gas-fired turbine in a CHP application may be more appropriate.

Desired resilience – The kinds of power sources you use also depend on the kinds of threats you face and how protected you want to be. We typically think of natural gas generators or fuel cells as highly resilient because natural gas pipelines are underground and not susceptible to transmission outages the way electric transmission lines are. However, natural gas is not immune to price spikes, fuel shortages, or even delivery interruption, as we saw in Texas in the winter of 2021. Communities that feel natural gas delivery is not reliable enough will likely want a technical strategy that employs more renewable energy such as solar, wind, biomass, and geothermal. The most resilient system will employ multiple generation types to diversify supply and reduce risk from any one kind of system failure.

Load types – The way your facilities use power also drives your technology choices. If demand is highly variable, with dramatic spikes during certain times of day, a battery storage system may be ideal to flatten out the load profile. If heating is a critical element of the base load, or more to the point, the emergency load during a grid disruption, then a CHP application that provides heat as a byproduct of power generation may make the most sense. If you have more interruptible loads that can be shut off for short periods as needed, you should consider how a microgrid design could support participation in demand response and load-shedding programs – adding an additional revenue stream to the system.

Available resources – Geography is clearly a driver when it comes choosing renewable resources, due to the availability of sun, wind, and other natural features that fuel renewable generation. We can also think about political geography as a driver since each state has slightly different regulatory and incentive structures. You should also consider your access to other fuel types and feed-stocks to help inform your decisions about which technologies are possible and which of those will be the most cost-effective and resilient.

On-Site Energy Technologies

For 100 years, most on-site generations consisted of diesel-burning backup generators – typically providing emergency backup power for a few hours or as much as a few days. The past decade has seen a rapid proliferation of alternatives, from campus-wide microgrids based on large-scale CHP systems to home battery backup systems like the Tesla Powerwall. It is now cost-effective in certain cases to run on-site generation systems year-round – realizing the economic and resilience benefits discussed throughout the book. However, nearly every rooftop solar system will go offline during a grid outage, to the surprise of some homeowners. If behind the meter DER did not go off-line during a grid outage, electricity would be sent back into the distribution system, where it could seriously endanger repair crews. The solution is to pair DER with software controls and a proper utility interconnection that allow the host institution to automatically "island" during a grid outage. As we discuss potential components below, remember that they will not yield resilience benefits unless the system they are a part of is explicitly designed to continue to operate when the utility feed shuts off.

It is helpful to first see the broad landscape: the most common distributed-energy technologies being used today. This book cannot present an exhaustive list of every energy system. Some technologies like heat pumps and solar thermal, for example, are not addressed because, though they can contribute to resilience, they do not produce electricity. Other technologies such as hydro turbines and geothermal electricity production are not addressed because they tend to only make sense at the utility scale or in special locations.

This section will likely be remedial for energy professionals with a strong working understanding of distributed-energy technologies. The descriptions below are brief summaries for readers who may only need a high-level understanding of technology options.

Thermoelectric Turbines

Any power plant that uses heat to create steam and turn a turbine is thermoelectric, including clean sources like nuclear and solar collectors. In this section, we are focused on combustion turbines – those that use a combustible fuel source and produce consistent base load power any time it is needed. The most common fuel source for these turbines is natural gas (methane). However, other fuels can be used such as diesel, gasoline, or biogas (a mixture of methane and CO_2 generated from organic material put through an anaerobic digester process).

These systems are dependent on the regular delivery of fuel, which limits their resilience. However, natural gas lines tend to be underground and are, therefore, much less vulnerable to (though not immune from) damage or interruption from inclement weather.

How It Works: First, a compressor draws air into the engine and pressurizes it. This highly pressurized air then enters the combustion chamber. In the chamber, fuel injectors fire the fuel source in to mix with the air. As this mixture is ignited, it boils water to create high temperature and high pressure. This steam picks up speed as it expands in the turbine section, rotating the blades of the turbine as it passes over them. Rotation from the turbine blades is used to generate power as well as to drive the compressor.

Where It Works Best: Thermoelectric turbines can work well in a variety of locations. They can cover larger base load applications with smaller footprints than solar PV or wind. However, they do require an external fuel source. From a resilience perspective, these systems work best in places with hardened natural gas supply. This means places where all gas distribution lines are underground and weatherized, and where the natural gas refinery feeding your distribution system is in an area with low risk of disruption from storms, flooding, earthquakes, or wildfires. If you have local access to methane from landfill gases or biogas from local farming operations, making use of them could make your turbine system both more resilient and more sustainable.

Pros:

- can provide large amounts of base load power;
- considered a mature and reliable technology;
- gas delivery infrastructure is more resilient than the electrical grid.

Cons:

- require an external fuel source (gas lines are not invulnerable), reducing the resilience factor;
- turbine systems are complex and highly technical, and require trained staff to maintain;
- associated with high initial costs and are ideal for larger applications;
- require a dedicated space to operate and may require the construction of a new facility;
- while efficient, these systems still use non-renewable fossil fuels, produce local air pollution, and contribute to climate change.

Fuel Cells

Fuel cells deliver constant, on-site power generation that can serve base loads while producing no harmful emissions. This technology uses hydrogen as a fuel and produces water as its only byproduct. Noriko Behling explains the technology this way: "Fuel cells are singularly remarkable in their potential for efficiently converting the energy locked up in chemical bonds to electrical energy. This efficiency is achieved because fuel cells convert the chemical energy contained in a fuel into electrical energy in a single step, extracting more useful energy from the same amount of fuel than any other known device."[46]

Of course, hydrogen cannot simply be mined or pumped out of the ground. It has to be produced using energy-intensive chemical processes. This means the efficiency and sustainability of fuel cells is determined by the hydrogen production process. Hydrogen is a colorless gas, but the industry refers to hydrogen fuel with a set of colors to denote its provenance. "Black" or "gray" are used to refer to hydrogen produced as byproducts of coal gasification – a process that has become increasingly rare as the abundance of domestic natural gas has grown. Today, the majority of hydrogen produced is "brown," or made through steam-methane-reforming process. This process uses steam and natural gas (methane) to isolate hydrogen molecules and produces a significant amount of CO_2 in the process. "Blue" hydrogen is a middle case – essentially the same as black, gray, or brown but paired with carbon capture technology to abate emissions. "Green" hydrogen is produced using clean electricity to power the electrolysis of water and is nearly emissions free.

How It Works: A fuel cell uses an electrochemical energy conversion process to turn hydrogen and oxygen into water while also generating electricity. In very simple terms, it does this by introducing a chemical catalyst to break electrons off the hydrogen and generate electricity. The now-ionized hydrogen atoms then join with oxygen atoms to form water. A single fuel cell produces a small amount of energy; so a number of them are stacked together to scale up the system to the desired output. There are different kinds of fuel cell technologies available today. Currently, the best technology for building or grid-scale applications is the solid oxide fuel cell. This technology is maturing rapidly, and new products are on the horizon.

Where It Works Best: Fuel cells work best in applications where base load power is needed, but there are concerns around emission, physical footprint, or noise. Some form of fuel is needed so that these systems work best in areas with lower costs of hydrogen or more common fuels like natural gas. Fuel cells can be installed with a CHP system so that they work well where there is a collocated thermal load. However, if generating heat is a primary goal, a fuel cell will generate much less than a thermoelectric turbine. Although fuel cells can be run at partial load, doing so results in less generation and therefore a longer payback from the system. For this reason, fuel cells are typically run at full capacity and should not be oversized beyond the requirements of the site.[47]

Pros:

- can provide emissions-free base load power (depending on how hydrogen fuel is produced);
- highly scalable technology;
- small footprint and clean, and quiet operation allows for deployment in even dense urban environments.

Cons:

- high capital costs;
- high maintenance costs;
- lack of hydrogen distribution infrastructure;
- the use of fossil fuels in the production of most hydrogen for fuel.

Combined Heat and Power

The goal in any exothermic power generation process is to use every bit of energy you can from the initial fuel and lose as little energy in the form of waste heat as possible. Both combustion turbines and fuel cells create waste heat as part of their energy generation process. Fuel cells are much more efficient, creating considerably less waste heat per unit of energy produced. Combined-cycle turbine systems use waste heat from the combustion process to increase system efficiency. CHP takes this process further by using the remaining heat in the system (after the combined cycle) to enable or increase the efficiency of other applications. The most common use of this heat is thermal comfort in nearby buildings. Using advanced absorption chillers, this heat can also be used to create cooling. In manufacturing settings, this heat may be used to increase the efficiency of an industrial process. CHP is a logical approach for distributed generation because the waste heat generated needs to be near energy end-users who can make immediate use of steam or hot air. CHP is a critical technology for any community installing gas turbine generation, and it should be considered for fuel cell applications as well.

How It Works: CHP can be thought of as an additional cycle in the combined-cycle turbine approach. After turning generation turbines, the hot water and steam that remains is run through a heat exchanger that can take the useful heat remaining and use it for space heating or industrial processes. The hot water or steam can also be pumped through insulated pipes to nearby buildings for similar uses there.

Where It Works Best: CHP works best where there is a thermal load that can be met using the system – particularly in colder climate areas. The earliest systems focused on industrial applications where process heating requirements were high, but the technology has proven to be effective in commercial and residential locations as well. CHP works best in these settings where it can feed heat to a district heating system that can distribute heat and steam to multiple buildings.

Pros:

- all the pros of combined-cycle gas turbines or fuel cells but with increased system efficiency;
- provides heat as well as power, further contributing to the resilience of institutions using these systems;
- can be purchased as pre-packaged system to reduce project complexity.

Cons:

- all of the cons of combined-cycle gas turbines or fuel cells with slightly more cost and complexity;
- requires infrastructure to transport hot water or steam between buildings to be effective across a portfolio;
- creating cooling with these systems requires expensive absorption chiller technology.

Photovoltaics

The advantages of solar PV for distributed resilience projects are many. PV panels can be placed easily on existing infrastructure such as a rooftop, on canopies over parking lots, or in adjacent fields. They produce electricity with no fuel, adding to their resilience value, and produce no emissions in the process.

Two factors that had limited the adoption of solar PV have receded over the past 10 years. First, the price has dropped precipitously. NREL found that from 2010 to 2020, there was a 64%, 69%, and 82% reduction in the cost of residential, commercial-rooftop, and utility-scale PV systems, respectively.[48] Second, the intermittency of solar PV generation (less power is produced on cloudy days, and no power is produced at night) has been partially addressed through the proliferation of solar plus storage systems. The hardware in energy storage systems has also seen significant price drops over the past decade making the solar-plus-storage model more accessible for a broader range of end-users.

How It Works: A PV cell is made up of semiconductors that absorb the energy from photons in sunlight. The photons strike the atoms in the semiconductor, freeing electrons. These free electrons are then directed by an electrical field to travel in one direction, creating current.

Where It Works Best: Photovoltaics work best where there is abundant sunshine, high electricity costs, and financial incentives for new installations. PV also is popular where there are favorable "net metering" laws, which are grid interconnection regulations requiring utilities to pay for solar power generated by an end-user and delivered to the grid. It is rare that all of these factors are in play in one location, but the technology has advanced to the point that just one or two of these factors will make the deployment of PV financially and technologically practical.

Pros:

- emissions-free power source;
- zero fuel costs;
- highly scalable;
- can be installed in a variety of locations (including building rooftops);
- low operations and maintenance costs.

Cons:

- intermittent power source requiring an energy-storage technology to sustainably meet base loads;
- large area of land (or rooftop) required to generate significant amounts of energy;
- PV panels produce direct current and require inverters to provide alternating current power.

Wind Power

After hydro power, wind power (in the form of a windmill) may be the oldest energy generation technology. Like solar, wind prices have dropped dramatically in the past decade. Lawrence Berkeley National Laboratory calculated that the cost of land based wind installations (those most relevant to local resilience) fell by 40% from 2010 to 2020.[49] Also like solar, wind is an intermittent power source, a problem that can be alleviated for resilience applications by matching it with energy storage. Depending on the type of facility, wind turbines can be hard to site, but new models of smaller scale rooftop wind generation are hitting the market all the time, and if there is field or ridgeline within reasonable distance of a facility or community, a full-scale wind project could still be integrated into a microgrid.

How It Works: In a wind turbine, wind turns the blades which turn a shaft connected to a generator. This creates direct-current power that must go through an inverter to provide alternating-current power. As a rule of thumb, the power generated by a turbine is roughly proportional to the square of its blade length. Generally speaking, doubling the length of the blades will quadruple the power output of the turbine. That relationship does not seem to bode well for the smaller wind turbines you would apply in a community or campus setting, but there is still a compelling case in certain locations. Traditional turbines with propeller-like blades are now being supplemented with vertical-axis turbines – some of which are much more suitable for urban environments.

Where It Works Best: Wind turbines work best where there are strong, consistent winds, high electricity costs, financial incentives for new installations, and net-metering laws. At the community scale, you should think about locating turbines away from large buildings – in large open fields or on ridgelines. For denser urban environments, architects and engineers have been increasingly finding ways to use small turbines to harness air flows created by buildings and even passing traffic.

Pros:

- emissions-free power source;
- zero fuel costs;
- turbines take up very little space and can easily share usage of the land on which they are sited.

Cons:

- wind is an intermittent power source requiring an energy storage technology in order to substantially address base loads;
- turbines need to be engineered to withstand specific extreme weather threats, including extremely high winds or freezing temperatures;

- turbines are highly visible and can create noise and visual distraction from natural environment – characteristics that nearby residents may find objectionable.
- wind turbines produce direct current and require inverters to provide alternating current.

Energy Storage

The purpose of energy storage, in grid-connected or microgrid applications, is to increase stability of our energy systems. There are certain times when we demand more energy than other times. For example, winter mornings, when buildings must be heated for the coming day, summer afternoons when the sun is strongest and the most cooling is needed, and the 5–7 p.m. period when office buildings and factories are still working and Americans are turning on the lights and appliances at home. All of these are times when energy demand rises. We call this instability "peak demand." Renewable technologies like wind and solar have variable supply. The wind blows sometimes and not others times. The sun shines sometimes and is hidden by clouds other times. It almost never shines at night.

Utilities often address demand instability with real-time pricing, or "time of use" rates – making it more expensive to use power during peak demand periods. Utilities also use demand charges (charges based on the highest level of demand in a particular period) to discourage spikes in demand. Supplying energy at those peak times is most expensive for utilities because they need to own and maintain generation equipment that may only be used for brief periods, often natural gas "peaker" plants. This equipment generates no income for the utility most of the year. Instability of demand is an expensive proposition for all parties.

If we can store energy when demand is low or supply is high and use that energy when demand increases or supply from renewables begins to drop off, we will reduce instability in the grid and increase the efficiency of our systems. From a distributed generation perspective, this means being able to supply more of our power with less on-site generation resources. The two most practical technologies to address these issues are batteries and thermal storage. There are other energy storage technologies such as fly wheels, pumped water, and compressed air, but these technologies are not going to be very broadly applicable for community-level energy storage needs; so they are not addressed here.

Battery Storage

Battery energy storage systems can manage power quality, allow for load shifting, and level the peaks and troughs of intermittent power sources. Batteries too have seen a steep decline in prices over the past 10 years, with Lithium-ion battery pack prices falling by 89% between 2010 and 2020.[50] In addition, many of the use cases that battery storage systems support help with payback, such as peak shifting and demand response.

How Battery Storage Works: All batteries rely on an electrochemical reaction between its two sides – the anode and the cathode. Oxidation on the anode side frees electrons that then pass through an electrolyte to the cathode side. A reduction reaction occurs on the cathode side where these electrons are absorbed. This process will continue until one of the electrodes runs out of a substance necessary to fuel the reaction. Charging of a battery is simply the reversal of this process.

There are two primary batteries for commercial use today on the building or campus scale: lead acid and lithium ion. Lead acid batteries are bigger and heavier. They are good for longer-term base load power to operate your systems where you do not need a high degree of variability in power. Lithium-ion batteries are a newer technology. They can be much lighter and can be scaled down in size (hence their use in cell phones). They can tolerate a wider temperature range and last longer than lead acid. These batteries are typically more expensive but can discharge their energy quickly to improve power quality, support utility grids, and play a more dynamic role in demand response.

Where It Works Best: Batteries work best where they support other technologies (like wind and solar) or utility demand programs. The ability of batteries to turn intermittent generation into an energy system that can cover base loads is a highly valuable application.

Pros:

- energy storage supports the use of intermittent power sources;
- can support demand response programs and reduce your peak load;
- can help manage power quality issues on site;
- extends backup power time during a grid outage.

Cons:

- even with decreases in price, these systems can still be expensive;
- batteries require space and electrical and physical infrastructure.

Thermal Storage

Thermal storage refers to a wide array of technologies that support the use of excess thermal energy some period after it was created. Thermal storage can be used for both heating and cooling, using storage media including water, earth, bedrock, or eutectic chemical solutions. The most common application of this technology for the institutions discussed in this chapter is the creation of ice (using a chiller system) during off-peak hours for use in cooling during peak hours. This is not a technology that produces or stores electricity. However, the impact of this kind of thermal storage on energy resilience (particularly in warmer climates) is significant. Thermal storage allows clients to install smaller chiller systems in their buildings. Those smaller chillers, operating at 100% with supplemental chilling from the ice storage, make for a much more efficient system than having a large chiller running at partial load. Alternately, some facilities are using thermal storage for demand response – waiting for a curtailment request from a utility and then shutting down all chillers, drawing all cooling from their ice system. Whatever scheme is used, thermal energy storage can be very effective in reducing costs and increasing resilience in environments with a high level of cooling-degree-days.

How Thermal Storage Works: The thermal storage application we are focusing on here is relatively simple. It uses your chiller to cool a transfer medium (a water/chemical mixture or glycol) that then travels through a heat exchanger in a tank (or tanks) of water. The water is slowly turned to ice. When additional cooling is needed for the building, the transfer medium is run through a separate heat exchanger in the HVAC system where fans blow air over the cold pipes, creating cooled, conditioned air.

Where It Works Best: Thermal storage is essentially a load shifting technology; so the highest paybacks for these systems will be in areas with high demand charges and expensive real-time pricing. These systems work best in large buildings with high cooling loads – particularly in warm weather climates.

Pros:

- can support demand response programs and reduce your peak load;
- allows most users to reduce the size and increase the efficiency of their chiller systems;
- creates a useful load for on-site generation during off-peak hours;
- creates cooling with very little energy during a grid outage.

Cons:

- not as applicable in colder climates;
- water tanks require space and plumbing infrastructure;
- there is some efficiency loss in the thermal storage process.

Microgrids

A microgrid is the stage upon which the energy generation and storage technologies discussed in this chapter can perform. In order for any of these technologies to realize their potential for improving energy resilience, they must be capable of continuing to power their host facility when the utility grid goes down. As we have discussed, on-site generation sources such as rooftop solar are designed to shut off when the grid goes down to protect line workers. However, with the right interconnection with the grid, an automatic transfer switch can be used to transition the on-site system into "island mode," a term used to describe continued operation, separate from the utility grid. This may be the most important quality that defines a microgrid, but it is not the only one. A microgrid must manage power across multiple loads and often among multiple DERs. This functionality requires some kind of control system and software, from rudimentary to highly advanced and autonomous.

Three main drivers have been behind nearly all microgrid deployments over the past decade: managing the cost of energy, sustainability, and of course, reliability and resilience. The right microgrid control system – some combination of meters and other sensors, PLCs, and SCADA software – can maximize each of these benefits and balance them based on the needs of the host facility.

A microgrid can manage cost in a variety of ways. It can actively favor on-site energy resources over the grid at times of peak demand to cut down on demand charges. If a facility is subject to time-of-use rates, the microgrid can time on-site production to coincide with higher rates and time the charging of energy storage to take advantage of lower rates. A microgrid can support participation in demand response programs, allowing even large facilities to scale down demand when asked by ramping up on-site production. A microgrid can also be designed to actively select the lower cost source of energy between on-site generation and the power grid. If, for instance, the cost of electricity is high and natural gas is low, the microgrid can favor on-site operation of a CHP plant. If natural gas prices soar, the microgrid can draw more from the electric grid. Finally, if a microgrid is built using an Energy-as-a-Service arrangement, a host institution can work with the microgrid developer to set a predetermined rate structure for the energy it will produce, thus shielding the institution from energy price fluctuations.

All of these cost saving functions improve the value proposition for on-site clean energy resources. As we have discussed, solar PV, CHP, fuel cells, wind, and storage are all well suited to on-site deployment. Payback periods have come down on all of these as the cost of the technology has fallen, but adding in the grid interactive capability of a microgrid (peak shaving, demand response, etc.) shortens the payback period even more. Add to this the resilience benefit that microgrids offer, and the business case for on-site renewables can start to look undeniable. In addition to simply enhancing the value of on-site renewables, a microgrid can allow its host institution to actively manage various energy sources to maximize resilience. Theoretically, a microgrid control system could be used to favor grid power at times when the grid mix is the greenest – on sunny, windy days for instance.

The ability to island is at the heart of the resilience value of a microgrid, and it may be the single most potent resilience tool discussed in this entire book. However, there are also other ways in which microgrids support reliability and resilience, even during normal operation. Since the DERs that make up a microgrid are always running, they can act as a primary source of power for an institution, with the grid as a backup. If any piece of equipment within a microgrid fails, controls can be configured to direct other resources and the grid to step in to fill the void. Energy storage technologies – electrical or thermal – act as a backup for both the grid and on-site resources. If there are certain loads in a facility that are particularly critical, such as life-support, the microgrid can direct them to draw energy from an on-site CHP system first, then the grid, and then on-site storage, essentially establishing three layers of resilience.

How It Works

To understand microgrids, it helps to first think about the macrogrid. The utility grid features large power plants generating power all the time and smaller "peaker" plants that start up when the demand becomes too great. Long transmission lines and distribution networks to get the power to end-users. A central control system manages voltages and attempts to balance the load across the system. Except for transmission, a microgrid has all of these components, on a smaller scale.

The components of a microgrid typically include the following:

- Transfer switches – These are the interconnections between the local microgrid and the power distribution from the utility grid. These systems are more than just a simple switch. They must match power frequency with the grid feed so that there will not be power quality problems with disconnecting or reconnecting with the grid.
- Generation sources – Some combination of on-site generation and storage resources are selected based on the host institutions' particular needs and opportunities, and scaled to power just the loads connected to the microgrid.
- Energy storage – As discussed in the section on battery storage above, these systems are critical for improving power quality and leveling out the load provided by intermittent power sources.
- Inverters – Almost all energy loads in the United States operate on alternating current. Direct current power from generation sources like solar panels needs to be converted into alternating current before it can be used.
- Microgrid control system – Just as the utility has a grid control system, a microgrid needs a control system as well. This combination of hardware and software enables the monitoring of loads, generation sources, and storage. It coordinates the flow of power within the system. In advanced systems, control systems can determine which base loads to connect or disconnect from the microgrid based on available power from generation sources. This infrastructure ensures that whatever power is produced is delivered to the most critical loads with sufficient power quality.
- Distribution lines – Connecting local generation sources to multiple buildings frequently requires the addition of power connections between those facilities. For communities that already connect to the grid in discrete ways (like military bases or college campuses), this can be simple. For communities integrated into the power grid in a more decentralized way (like local governments), this connection is more complex and will likely require the addition of new power

lines. The installation of lines may require regulatory approval and engagement with the local utility.

Where Microgrids Make Sense

Microgrids are ideally suited for campuses of adjacent buildings, or in the context of a municipality, a group of buildings in close enough proximity to facilitate connection on electrical and IT networks. Other criteria that make an institution particularly well suited for microgrid adoption include the following:

- a mission with high up-time requirements, as with hospitals, data centers, and all of the other cases described in this chapter;
- availability of financial incentives that reduce payback periods for the installation of DER and control systems;
- a regulatory and utility environment that supports distributed generation and periodic disconnection from the grid during an outage event.

As the challenges discussed in Chapter 1 (problems with grid reliability, extreme weather events, and energy costs) increase, microgrids will make

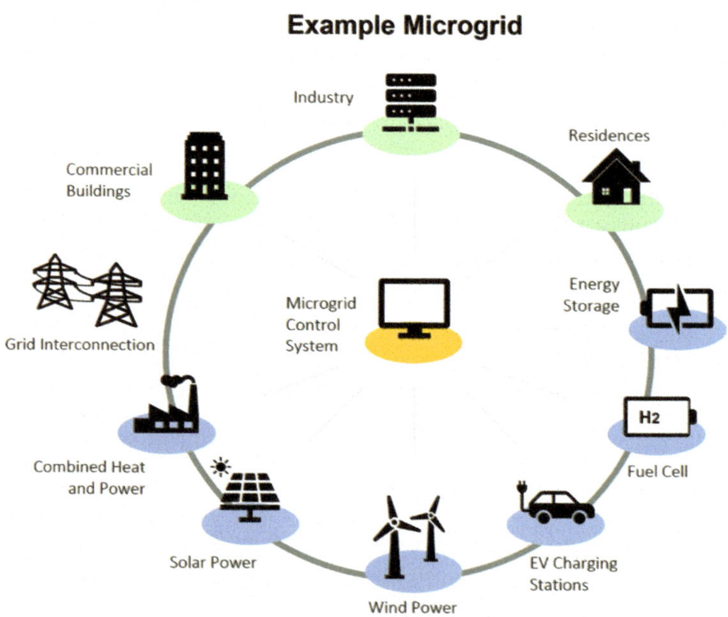

Figure 6.2 Example of a microgrid system.

sense in a broader array of scenarios. This book focuses on U.S.-based applications for microgrids. However, microgrids are also an ideal solution for electrifying remote areas, islands, and developing nations – where deploying microgrids may be much more cost-effective than building long-range transmission lines.

Local Energy Resilience Maturity Model

	No Program	**Reactive Program**	**Managed Program**	**Proactive Program**	**Best-In-Class Program**
Stakeholder Engagement	Energy resilience is the purview of one energy manager, with no oversight or collaboration.	Designated energy manager engages leaders across a variety of departments on energy management, mostly to inform them of new policies and projects.	Leadership engaged in the energy planning process to guide and buy-in to new initiatives. Regular updates are delivered to various department heads.	Designated team of stakeholders to create an energy resilience plan and assume responsibility for executing it, including energy managers, executives, and finance.	External stakeholders are also engaged to give input in the energy resilience planning process. Energy resilience plans and annual progress are reported on publicly.
Goals	No goals around energy.	A general organizational focus on reducing energy waste where possible.	A goal to reduce energy consumption by a certain amount by a certain date.	Defined goals to improve resilience, often including a goal to increase utility management and on-site energy generation by a certain amount and date.	Deadline-based, public goals that address energy efficiency, on-site generation, and operational aspects of documented resilience program. Energy goals are stated as part of the financial goals of the organization.

Strategy	No strategy for energy performance.	Energy management is assigned to a certain party, but no strategy is documented.	An energy strategy is written at the Energy Manager/ VP level and shared with facility management staff. The strategy focuses on short-term, quick-payback projects.	Energy strategy is reviewed periodically to ensure alignment between tactics and goals, and that progress toward goals is on track. The plan is documented and signed-off on by leadership.	A team of stakeholders draft a multi-year energy roadmap addressing all aspects of the energy resilience. Roles are well defined, tactics are aligned with stated goals, and there is strong support from upper management.
Energy Efficiency	No consideration is made for energy efficiency technologies. Lowest first-cost technologies purchased.	Replacement of end-of-life technologies with more efficient models. Technologies are installed without addressing systemic building issues.	Energy efficiency considered in facility planning. Cost–benefit analysis (CBA) used to prioritize technologies with quick payback.	Energy audits of low-performing facilities to identify efficiency opportunities. Projects with resilience benefits are prioritized, and energy savings are tracked.	Energy audits performed at all facilities, and retrofits are evaluated using lifecycle costing methods. Performance is tracked using building energy models and software.
On-site Generation	No backup power available.	Limited to some emergency diesel backup power at certain locations or for particular loads.	Energy assessments to prioritize loads and evaluate needs for additional backup power given increased risk of outage.	Installation of new on-site generation and microgrid controls to cover certain facilities or loads during a grid outage.	Advanced microgrid fed by clean and resilient on-site energy, matched with storage and advanced controls capable of sophisticated grid interaction to optimize cost and sustainability performance.

Conclusion

Grant Ervin, the Chief Resilience Offer of the City of Pittsburgh, explained his approach to energy resilience to us this way:

"We look at energy resilience from two aspects, our own operations and our effect on the city at large. In terms of our own operations, we are like a small to mid-sized company with a few thousand employees, a fleet of vehicles, and a portfolio of facilities. We developed our climate and resilience plans to leverage our control over these assets. We also understand that government actions can have impacts on operational and social resilience throughout the city. When you bring these things together, you can create a sharper picture of where you can make investment and policy decisions that support resilience for everyone in Pittsburgh."[51]

As Ervin astutely observes, everything a municipality does to protect the energy resilience of its own operations also protects all the residents who rely on city services, from fire and police to healthcare, water, and even electricity itself. The same principle applies to a college campus, hospital, and even a military base. The impacts of energy interruptions are felt first and foremost at the local level, and local intuitions have the most capability, and perhaps the most responsibility, to address them. That said, the most effective local energy resilience strategy is carried out in concert with regulators, as we discussed in the previous chapter, and utilities, as we will discuss in the next chapter.

7

Government Resilience: Policy and Programs

The government's role in addressing energy resilience includes two primary functions. The first is providing public goods like national defense, education, and public health. The second is addressing what economists call market failures. This could take the form of regulating natural monopolies (like electric utilities), performing high-risk research and development that the market would not normally pursue (as we do at our national laboratories), or ensuring that externalities (like climate-change-inducing carbon emissions) are built into market prices or otherwise addressed.

The delivery of public goods and services requires access to resilient, reliable energy. As with businesses, most government services simply cannot be delivered without energy. The difference is that government services are often life-saving activities that are critical to the population as a whole. For this reason, government entities have been at the forefront of energy resilience planning. The U.S. military, for example, has been an early adopter of microgrids, installing systems on bases which allow them to island themselves and provide on-site energy independent from the main grid when necessary. In October 2021, the White House released plans developed by more than 20 federal agencies which outline the steps each agency will take to ensure that their facilities and operations adapt to and are increasingly resilient to climate change impacts.[1] When planning for energy resilience, governments usually regard their own continuous operations to be a critical goal.

To address market failures, governments can support or mandate industry activities that will result in improved energy resilience. This is a complex process both because of the numerous channels for addressing these failures and because the responsibility for addressing these issues is spread broadly among federal agencies and state governments. Having a broad, overarching energy resilience strategy that is incorporated into the planning of multiple government agencies is the key to leveraging government power. Without coordination, governments risk wasting time on small, siloed efforts that have minimal impact and fail to leverage each other.

Government support, be it technical or financial, has been – and will continue to be – critical to fostering energy resilience. These efforts can focus on commercializing new technologies, delivering technical support and education, or simply providing targeted funding opportunities to make up for market gaps in financing of projects that have societal benefits. This work is not done by just a single entity. The U.S. Department of Energy may take the lead at the federal level, but other agencies are also doing critical energy resilience work. For example, the Departments of Defense and Homeland Security both work with industry partners to share their lessons learned and foster resilient energy systems around their facilities. Many states have their own agencies and centers of excellence that support energy resilience efforts through research, partnership, and grants.

This chapter will discuss how government policies and programs can foster energy resilience. Our primary focus will be federal and state government activities. While city governments are highly active here, they have unique opportunities and challenges and are addressed in more detail in Chapter 6.

Government Actions in Support of Government Energy Resilience

Much like businesses and communities, governments hoping to improve their energy resilience should take a portfolio-wide approach. In some cases, these efforts can be driven by legislation, creating a legal mandate for action and/or granting the government new powers to address its own resilience. More often, these efforts are driven by executive leadership (presidential or gubernatorial) with mandates for action handed down to agency heads. These efforts typically utilize the same best practices identified in this book. A thorough resilience planning process (like those from the Department of Energy and Department of Homeland Security described in Chapter 3) is vital to providing direction to a government agency pursuing resilience. It can also help define a progress report for an agency's leadership or legislative group.

Evaluating Risks to Government Facilities

Government agencies are likely to already have an internal analysis of which activities are mission critical and which are most susceptible to various threats. Understanding the acute nature of energy risk may involve going one or two levels deep with team leads to understand not only how an energy loss would affect their direct operations but also how it would impact their supply chain and the network of other government and private functions on which they relay. Given the complex and interrelated nature of government agencies, a power outage in one part of the country could easily impact government activities in another part.

To better demonstrate risk identification, following are some examples of resilience risks we might see at different levels of government.

Federal government: In Afghanistan, long supply chains for diesel fuel for front-line positions created vulnerability for U.S. military operations. This presented both an operational security and an energy resilience risk. The military responded by developing more advanced, easily deployable photovoltaic panels to reduce the load on generators at forward operating bases.

State government: A state government, having experienced multiple storm outages, realized that reduced energy usage would considerably improve its ability to respond to outages. The state instituted a holistic review of its own energy use, and it identified opportunities to improve energy efficiency and on-site generation and energy storage.

Local government: A local government, which had emergency backup power at all their facilities, realized that the local hospital only had limited capability for 72 hours during an outage. Government officials worked with the hospital to explore the potential of a community microgrid, centered on the hospital, which could provide additional resilience.

Recognizing the Challenges

Once a government organization understands the risks it faces, it can consider possible solutions. Governments tend to have a greater level of control over the land and facilities where their operations take place, giving them options that may not be available to some commercial players. Governments also have a vastly different mandate than commercial entities. Maintaining government functions like emergency communications, police, fire, and even wastewater treatment requires a different set of calculations when it comes to the value of resilience.

When asked about the biggest challenges associated with implementing resilience programs, Crystall Merlino, Director of Resilience, Sustainability, and Energy Management at the Department of Homeland Security, identified funding as the primary challenge, particularly due to the heavy technology and infrastructure needs for most resilience programs. She also acknowledged that unlimited funding did not ensure success.[2] Indeed, funding resilience efforts at a government agency can be much more complex than a similar effort in the private sector.

Securing the funding faces a number of hurdles, beginning with the fact that any effort requires a multi-year commitment. Plus, large programs demand at least a year of internal Congressional approvals before they can be added to a budget request. State agencies go through a similar series of budget approval steps. And even if the proposed solution can fit within an agency's existing budget, competing programs and efforts can supplant a project as seemingly nebulous as "energy resilience." Agency staff and leaders may see focusing on resilience as yet another unfunded mandate that is tangential to their core mission. While energy resilience can be critical for accomplishing that mission, tight budgets can make it difficult to justify expenditures outside of traditionally mission-oriented activities. Jennifer DeCesaro, the Director of Recovery and Resilience at U.S. Department of Energy, stressed the importance of communication in securing funding – ensuring that other parts of the organization understand the critical role that resilience can play in mission readiness. Demonstrating the value proposition over time seems to be the primary challenge and most likely avenue to success."[3]

Government agencies face the complexity of obtaining goods and services through intricate government purchasing channels. While private companies can issue a contract directly to the firm of their choice or possibly issue a solicitation and select the best proposal, government entities have additional rules complicating and lengthening their buying process. Every contract must be (in some way) open for public bidding. There are rules around how much government spending must be sent to small, disadvantaged businesses. Government agencies are often pushed to select the lowest bidder over the optimal solution. Finally, the timeline of issuing a solicitation to the start of work can be drawn out to a potentially unacceptable degree.

Government agencies may operate across a wide geographic range, complicating the sourcing of solutions. The right resilience solution in New York will likely be different than the right solution in Louisiana. Jennifer DeCesaro told us about the Department of Energy's approach to helping governments and private entities to meet this challenge. "We take a technology-agnostic approach and focus on process more than driving specific solution pathways. The right solution will be unique to each community based on the resources they have available and their priorities as a community. We provide a framework for how they go about planning their customized energy transition. We have tools that help them to evaluate resilience investments against each other – developing this in partnership with two utilities."[3]

One of the earliest and most ambitious government efforts to address energy resilience was the U.S. Department of Defense's Smart Power Infrastructure Demonstration for Energy Reliability and Security (SPIDERS) program. This effort started in 2013 and was focused on deploying cybersecure microgrids at U.S. military installations. Leveraging the expertise of multiple national laboratories and the Department of Energy, the four-year DOD program had four primary goals:[4]

1. protect critical defense infrastructure from loss of power because of physical disruptions or cyberattack to the bulk electric grid;
2. sustain critical operations during prolonged utility power outages;
3. integrate renewable energy sources, energy storage, and other distributed generation to power defense critical infrastructure in times of emergency;
4. manage DOD installation electrical power and consumption efficiently to reduce petroleum demand, carbon "bootprint," and cost.

While the SPIDERS program had some challenges in the areas of transparency and commissioning, the overall project achieved what it set out to do: prove the efficacy of microgrids to ensure the operational readiness of military bases in grid outage situations. While the SPIDERS program is complete, the military continues to expand its use of on-base microgrids around the world.

Government Actions to Support Market Energy Resilience

For government entities to affect energy resilience at a societal level, they must go beyond their own operations and use the levers of government to support much broader market adoption of solutions. Governmental entities have four primary means by which to impact the energy resilience of non-governmental activities and facilities:

1. research and development that characterizes risk or advances energy resilience solutions;
2. support for energy resilience efforts through technical support, grants, or tax incentives;

3. driving market activities through their own purchases and policies;
4. regulatory changes that mandate or incentivize energy resilience of buildings or the energy grid.

The most effective way for government to move the market is to use all four of these approaches since they tend to impact different aspects of the market. These four areas of support can also be designed to complement each other. For example, government lab research into energy storage technologies could be commercialized with the help of a government grant program and purchasing requirements. Regulatory direction could ensure that the electric utility sector creates a pathway for connection of those systems to the grid. Figure 7.1 shows the characteristics of solution commercialization.

New York State's Reforming the Energy Vision effort (NYRev) is an excellent example of how a government can leverage its entire suite of tools to address energy issues. This effort, led by Governor Andrew Cuomo, was woven through 40 programs across five different state agencies as well as the New York Public Service Commission. It sought to direct state government efforts to create an energy system that produced fewer greenhouse gas emissions, lowered energy consumption, transformed the transportation sector, and improved the resilience of the energy system – all while bolstering a clean energy economy to bring new jobs into the state. State agencies incorporated the NYRev goals into their strategic plans and activated funding sources for grants and program funding. A Clean Energy Fund was created to attract more private capital into New York's emerging clean energy economy, accelerate the adoption of energy efficiency, and

Stage	**Early Stage R&D**	**Pilot Project/Commercialization The "Valley of Death"**	**Market Adoption**
Characteristics	▪ High-risk R&D ▪ Early-stage technologies ▪ Proofs of concepts	▪ Pilot projects ▪ Data collection ▪ Business model creation	▪ Clearing regulatory and market roadblocks ▪ Marketplace establishment ▪ Training/education
Government Support	▪ Laboratory funding ▪ Research grants ▪ Long-term financial incentives	▪ Agency programs ▪ Deployment grants ▪ Technology licensing	▪ Financial incentives ▪ Regulatory schemes that support transition ▪ Public/private partnerships

Figure 7.1 Characteristics of solution commercialization.

promote the adoption of renewable energy. The Public Service Commission enacted performance-based regulation to tie electric utility profits less to capital investment projects and more to outcomes like installation of clean energy, transportation electrification, and energy resilience.

The NYPrize program was one of the most noted energy-resilience-focused programs to come out of NYRev. This program, run by the New York State Energy Research and Development Authority (NYSERDA), was developed to support the increased deployment of microgrids in New York communities. The agency issued grants to 83 communities for microgrid feasibility studies. The program was successful in creating considerable interest around and research into community microgrids. The fact that few of these projects are still active provides useful lessons in the difficulty of making multi-stakeholder microgrid projects a success. A microgrid that serves both public and private entities with varied credit history and energy needs will always be difficult to organize, finance, and execute, but hopefully the many attempts that were made under the NYPrize program will help to support future government programs with the same aim.

Research and Development

While the private sector can be counted on to develop and innovate in technologies with clear commercial potential, companies rely in many ways on government R&D. Often termed "high-risk R&D," these activities play a crucial role in advancing energy solutions. This is because government entities can pursue areas of research without concern for immediate commercialization of the outcome. This allows them to work in areas of basic science (that will one day lead to other developments) and on a broad variety of technologies before a clear market winner emerges. A great example of this government-led research has been the work done on renewables and energy storage at national labs. Without lab efforts on this front beginning in the 1970s, the global markets for these technologies would be considerably farther behind where they are today.

Government entities can maximize value of government R&D dollars by investing in concert with private entities. This can take the form of public–private partnerships, in which the workload and benefits of R&D are shared. We have seen this model work quite successfully with organizations like the National Renewable Energy Laboratory (NREL), which partnered with manufacturers of renewable energy equipment for research and equipment testing. Industry experts can also advise government agencies and labs on the industry's weak points and needs, which can lead to better target government R&D funding.

There is, however, a risk of overtaxing industry experts for such projects, as Sharla Artz of Xcel Energy found to be the case in the cybersecurity sector: "There are so many government activities and working groups focused on best practices for cyber security. This is too much to ask of experts. They need to streamline this. If we do the same thing with supply chain that we did with cyber, our few experts will not be utilized well to solve the problem. Government needs better information on what's being done in the field before they start leading or even coordinating. Creating a working group for the sake of a working group is a waste of time."[5]

To advance resilience, government funding can be deployed to develop technical solutions in the areas of distributed energy resources and advanced microgrids. Alternatively, government R&D funding could target more market-centric concepts like system controls, resilience metrics, or business model development. A healthy government R&D effort will continue to play an important role in moving energy resilience from basic backup generation to advanced systems that work in harmony with people, technology, and market forces.

The work currently underway at the National Renewable Energy Laboratory (NREL) provides an excellent example of publicly funded resilience R&D. NREL scientists and analysts use Department of Energy funding to advance areas of resilience including:

- quantifying the value of resilience – building quantitative models to show communities, utilities, and policy makers the benefits of investing in resilience strategies for the grid;
- developing the North American Energy Resilience Model which will allow users to predict the impact of threats, evaluate, and identify effective mitigation strategies, and provide support for black-start planning to enhance energy and economic security;
- conducting general science and direct application research on resilient energy technologies like energy storage and microgrids;
- pursuing direct partnership with the U.S. military to identify resilience gaps and highlight opportunities for on-site solutions, such as solar PV paired with energy storage and diesel generators, to improve resilience within critical loads;
- forging broad partnerships with industry and other agencies (like FEMA) to evaluate resilience challenges and provide technical assistance for resilience planning efforts.[6]

Government Programs: Technical Support, Grants, and Tax Incentives

Once technology has developed to the point that it has commercial potential, it must cross what is often referred to as the "valley of death." The transition from laboratory to commercial sales is challenging even for a viable technology because it will suffer from being new, unknown, and lacking a long track record of success. Establishing new technologies (in energy or any other market) may require large private firms to make major investments. This is especially true in the energy industry where new solutions are often both highly regulated and capital-intensive.

Government support can play an important role in moving the market and helping technologies establish a beach head for commercial success. Sometimes, this can be done with a narrow focus on one aspect of the market that is difficult for private companies to move, such as public education around the safety/efficacy of a new technology or funding early pilots to prove out business cases. More often, a blend of government support is needed. This is what NREL discovered when it evaluated optimal approaches for energy technology commercialization: "Supporters of new clean energy development hypothesized that commercialization of hardware-centric energy solutions could be accelerated with 'enabling ecosystems'—more 'organic' arrangements combining strong technical support, business/product

Figure 7.2 A lab in the NREL ESIF facility simulating energy usage across multiple buildings. "Energy-Efficient and Grid-Interactive Buildings." National Renewable Energy Laboratory. 2022.

verification, flexible financing, and commercial validation. In aggregate, these components are expected to shorten technology development cycles."[7]

Examples of government support for the commercialization of energy technologies are plentiful. For instance, the EPA's Green Lights program of the early 1990s, which would become the ENERGY STAR program, was a technical support and labeling program responsible for dramatic increases in the efficiency of a wide array of appliances. In another example, the Department of Energy's Energy Efficiency and Renewable Energy Office has provided market education and regulatory support for renewable power sources. Most recently, we can look at the impact that tax credits have had in accelerating the renewable energy and electric vehicle markets. For example, the U.S. production tax credit (PTC) for renewable energy has had a huge impact on the development of renewables – particularly wind power. Federal and state tax credits have also helped accelerate the adoption of electric vehicles. Finally, programs like state and federal "Green Banks," (which provide financing to sustainable investments) and loan guarantees (which provide federal backing for loans in the public interest) have both been used to remove financing barriers to projects in the energy space.

In addition to these ongoing government programs, Congress has occasionally established funding mechanisms that have led to major progress in energy resilience. One example of this was the 2009 American Reinvestment and Recovery Act. This act provided $787 billion (later raised to $831 billion) to spur economic movement out of the Great Recession of 2008. The bill included over $120 billion in infrastructure projects, and many of them focused on energy resources. One of the larger programs, the Smart Grid Investment Grant (SGIG), allowed the Department of Energy to direct roughly $9.5 billion to 99 cost-shared projects involving more than 200 participating electric utilities and other stakeholders. These projects were designed to modernize the electric grid, strengthen cybersecurity, improve interoperability, and collect an unprecedented level of data on smart grid operations and benefits.[8]

In 2021, President Biden signed another large infrastructure bill. The Infrastructure Investment and Jobs Act (IIJA) will provide $1.2 trillion in spending, $550 billion of which will be allocated over the next five years. The bill included the following provisions related to energy resilience, among others:

- $73 billion for the electricity grid – Upgrades to the country's power systems that will improve communications, resilience, and connections for renewable energy.

- $47 billion for climate resiliency – New funding aimed at addressing climate threats like wildfires, coastal flooding, and extreme weather.
- $7.5 billion for electric vehicles – Helping to commercialize EVs by accelerating the deployment of charging stations nationwide.

Government Purchasing

Efforts by the government to ensure its own resilience can have a wide ripple effect. If two states want to improve resilience, one might do it through market partnership programs and the other might focus on resilience of state government facilities. Other states can then benefit by analyzing the success of these programs, determining which was easier, which was more expensive, and which had the greatest impact on statewide resilience. Whether the lessons come quickly or slowly, states and countries will always be watching each other to determine how the experience of others can impact their own decisions.

Very often, decisions or requirements of one government or agency prepare the market to adopt those same requirements more broadly. A famous example of this is California vehicle emissions standards. Since the 1960s, the state's regulations have been more stringent than federal standards. Given the size of California, auto makers were forced to develop technology that allows them to sell more efficient cars in that state. The result was the availability of automotive technology that was cleaner and more efficient – technology that was usually just incorporated into the standard vehicle offering across all 50 states. When the federal government institutes a new requirement for their purchasing, any company that supplies a high volume of goods to the government will soon adopt this requirement. It may not be a regulation or law, but if your largest customer simply will not buy from you unless you offer a certain feature, that feature becomes standard quite quickly.

Finally, there is the power of influence. One government's policies can often influence another's because of hierarchy (federal laws or spending decisions influencing state behavior), treaty (on issues like trade or climate), or public opinion. For example, if the federal government decides to support microgrid development, states may establish microgrid programs allowing them to leverage that federal spending to the fullest. If the U.S. and Canada are negotiating trade issues, one party may want an energy-related provision included to support their goals on that front. If one state is taking advantage of offshore wind resources, it may create public pressure for neighboring coastal states to evaluate this as well.

In all of these ways, the efforts of a single government entity can have concrete impacts on the goals and efforts of other entities and the market at large. When governments want to be even more proactive, they can directly support market energy resilience.

Modernizing Regulatory Frameworks

Regulation is an important tool for steering the course of the American energy system – primarily through the control of electric utilities. From the 1930s to the 1970s, regulators at the state and federal level created mandates and incentives to induce electric utilities to connect every part of the country to the grid. Even when running power lines was expensive and would serve few customers, utilities were compelled to expand because universal electrification and service expansion were the primary regulatory drivers of the day. For this reason, a "cost-of-service" regulatory framework was adopted for electric utility remuneration. Under this approach, utilities are paid back for their expenses and make a fixed rate of return on all infrastructure investments in the grid (power lines, transformers, switchgear, etc.). Although this incentivized utility investment in the grid, it did not address other desirable outcomes.

In the 1970s, with the nation largely electrified, regulation started to respond to more diverse interests and push the industry to both lower costs and mitigate its own market power. The Public Utility Regulatory Policies Act of 1978 required electric utilities to buy power from independent generators, creating the first crack in the vertical integration of utility monopolies. The newly created Environmental Protection Agency and Department of Energy deployed programs aimed at encouraging energy efficiency and reducing air pollution.

The 1990s saw the rise of "utility restructuring," an effort in roughly half of U.S. states to force utilities to completely divest their generation assets. The hope was that by separating the generation side from the transmission and distribution side, regulators could create open market competition between generators while focusing electric utilities on the job of investing in and managing their grids.

Today, we are seeing yet another shift in regulation in the U.S. and other countries. Regulators are looking for ways to address a variety of challenges in their effort to modernize an aging grid. The regulatory focus, in fact, has expanded to include the balancing of intermittent renewables, addressing new demands related to electric vehicles and data centers, coping with severe weather, and continuously improving system reliability. From an energy resilience perspective, there are some very specific steps regulators can take to spur market action, but effecting regulatory change is almost invariably a slow process. As Erik Svanholm of S&C Electric put it: "The rules were written with bright lines 50 years ago when there were no options. Now there are options and the laws simply haven't been updated."[9]

Clarity of Expectations Around Resilience

Energy resilience is a new-enough concept that most regulatory models have not yet incorporated metrics or standards around what is expected of utilities. Chapter 4 of this book discusses resilience metrics and how complicated they can be to select and implement. However, utilities are going to have a difficult time focusing on grid resilience without some kind of reporting metric to both help identify what strategies are most effective and to keep senior leadership focused on resilience as a driver of profit margin.

Once metrics have been established, whether they are quantitative or qualitative, regulators need to develop clear guidance around the levels of resilience expected from the grid. In most states, resilience still feels like a "nice to have" but not something for which utilities can justify major investments. Without establishing some baselines of and targets for performance, it will be difficult to generate innovation and movement to address major threats to the grid.

Remove Regulatory Barriers for Energy Resilience Projects Like Microgrids

Regulators can focus on identifying the regulatory barriers to (or sometimes simple lack of clear regulatory support for) projects that could improve the resilience of the grid. One of the starkest examples includes the ownership and operation of microgrids. Only a handful of states have developed clear rules around who can own microgrids, when and how a utility can rate case investment in a microgrid, and what the utility's responsibilities are for connecting behind-the-meter microgrids to the main grid.

When non-utility entities want to develop a microgrid, there are many questions that need to be answered. Developers often have little awareness of interconnection requirements and what the timing or cost will be of an interconnection study. Project leaders rarely have reliable information on the value of excess electricity that they might provide back to the grid. And projects that might cross streets or other rights of way with a power line could easily be vetoed by a utility claiming that the microgrid is now operating as a competitor and encroaching on the monopoly franchise of the local utility. All these issues limit and slow down the development of behind-the-meter microgrids.

When a utility wants to own a microgrid, the restructured market in about half of U.S. states means that those utilities cannot own the generation assets needed to power it. Even energy storage is treated as generation in many states, limiting utilities from deploying this highly useful grid resource. Utilities could sign an energy services agreement with a non-regulated company for the needed generation, but there are many states where it simply is not clear if the utility can do this – and, more importantly, whether they can incorporate those costs into their rate base for profit.

There are many critical questions to answer. If a utility can own a microgrid, under what circumstances will the regulator approve the investment in a rate case proceeding? Would only the customers directly served by the microgrid be tapped to finance the investment or would costs be shared among all the customers in the service territory? How does this change if the microgrid is serving a hospital or other critical infrastructure?

Answering these questions is complex but not impossible. Regulators can take proactive steps (as has been done in states like California and Hawaii) to establish a clear microgrid tariff that lays out when and how a utility can own microgrids and what obligations they have in connecting to behind-the-meter microgrids. Permitting processes can be simplified and expedited. Restructured states can create rules that allow utilities to bring microgrids into their networks – potentially classifying these systems as grid resources and not purely generation. In general, regulators need to carve out more resilience opportunities that both utilities and end-users alike can take part in by right and not only by special dispensation.

Multi-Year Rate Plans

Resilience is, by its very nature, a long-term investment. Traditional year-by-year rate making for electric utilities provides for long-term investments but, typically, only for well-understood, large-scale infrastructure projects like substation upgrades and switching equipment replacement. Moving from single-year rate plans to multi-year rate plans gives utilities the flexibility to make more forward-looking resilience investments with time to recoup the upfront capital expenditure. It is also important to include flexibility within a multi-year period to adapt to changing conditions (both in terms of threats to the system and available solutions).

Performance-Based Regulation

Most regulatory systems are not designed to engender the kind of investments needed to improve the resilience of our energy systems. The present utility regulatory environment in many countries involves some minimum standards of service, but it primarily focuses on cost-of-service (COS) regulation, where utilities make their profit based on their investment in grid infrastructure. You may think, "That's fine, we need new infrastructure!" Unfortunately, traditional COS regulation does not easily allow for transformational resilience investments, nor does it reward solutions that are not hardware-based. It rewards utilities for building the same thing they have always built, the way they have always built it. Regulators have traditionally been focused on short-term efficiency and controlling rates rather than the future needs of customers. This can make the grid of the future a hard-sell.

Utilities need three things to address the changing energy landscape:

1. a thorough understanding of new energy solutions;
2. the ability to include these investments in their spending plans;
3. a regulatory framework that makes modernizing the grid the most economically attractive pathway.

Regulators understand this, which is why some are seeking to transform electricity regulation. Progressive regulatory agencies like New York, Hawaii, and Illinois are supporting a transition from the traditional COS approach to a performance-based regulation (PBR) approach. PBR alters the profit incentive for utilities away from capital investment and toward meeting performance objectives and improved efficiency. The metrics used could cover anything but typically include reliability, resilience, customer satisfaction, safety, environmental protection, cybersecurity, deployment of advanced grid controls, demand-side management capability, and support for transportation electrification.

PBR can run afoul of the traditional utility/regulator compact that exists in some form in both regulated and restructured regulatory models. PBR also requires regulators to start picking winners and losers when it comes to certain outcomes and goals. Does society prioritize a clean grid or a reliable one? Does it want low rates or an evolution to a more advanced grid? These questions are being discussed by legislators and regulators while utilities and interest groups lobby for the answers they want. The keys to successful PBR, as has been seen in countries such as the UK, Australia, and Canada, are to:

- engage the utilities and market players early so that they can have input into the targeted outcomes, specific metrics to measure, and the feasibility of each goal;
- make performance metrics and incentives crystal clear to avoid confusion;
- avoid having too many metrics because they will dilute the impact of each;
- understand PBR as iterative and requiring feedback and changes in future periods.

Government Resilience Program Maturity Model

The maturity model below highlights increasingly ambitious activities that government entities could use to deploy a more robust energy resilience program across several areas. These maturity models can be a tool for evaluating your program's place on the spectrum of energy resilience or for setting targets in future efforts. When reviewing this, you can assume that any proactive steps in lower-tier programs are included in the higher-tier programs.

	No Program	**Reactive Program**	**Managed Program**	**Proactive Program**	**Best-In-Class Program**
Portfolio Resilience	Utilize the maturity model from Chapter 6				
Research & Development	Rely entirely on the market to advance technology.	Targeted research on technologies needed to achieve policy goals.	Basic science and targeted research related to policy goals.	Partnership with private industry to support cross-industry development and testing.	International partnership to advance basic science and technologies for resilience.
Commercialization	Leave commercialization up to private industry.	Education programs around technologies that match other policy goals.	Education programs and grant funding to advance new resilience technologies.	Education, grants, and technical support coupled with direct government purchase of resilience tech.	Collaboration across government agencies to demonstrate and normalize new tech.

Regulation	Cost-of-service regulation focused on traditional utility investments.	Establish clear guidelines on ownership and operation of resilience tech.	Establish/ track performance metrics for energy resilience.	Elimination of barriers for resilience efforts like microgrids and utility ownership of energy storage.	Performance-based regulation with resilience metrics tied to utility economic performance.
Purchasing and Supply Chain	Purchasing does not consider resilience.	Purchasing includes resilience considerations in purchase of energy equipment.	Evaluate resilience gaps in the supply chain and alter purchase requests to close those gaps.	Establish standard language in purchasing to ensure that all suppliers are meeting resilience planning best practices.	Active partnerships with suppliers to improve resilience throughout the supply chain and share best practices.

Conclusion

Whether addressing their own internal energy resilience or attempting to support resilience efforts in the market, governments need to think carefully about the multiplying effect their efforts can have. Government leadership in an area sends a message about what is prudent and where future requirements may arise. Federal, state, and local officials and regulators can drive progress if they recognize that these efforts demand a multi-year commitment as well as a thorough understanding of barriers – and how to surmount them.

Conclusion

As the threats to our energy system multiply and evolve, organizations are re-evaluating their resilience goals. This book is intended to break down the options and logical steps to advance energy resilience in the midst of a changing world. While we have presented strategies for three major organizational types (utilities, end-users, and governments) we hope that the reader will recognize that the implementation of energy resilience in any one of those is dependent on the other two. What is more, there are exciting opportunities for organizations at different levels to partner in advancing energy resilience across the entire economy. Threats to the energy resilience of one system are a threat to the resilience of all three. Lessons learned from energy resilience efforts in one area can and should be transferred to the others.

An excellent example of this kind of cross-sector implementation of energy resilience concepts can be seen in the recent history of California. The drought conditions that have plagued California over the past two decades have created serious challenges for the generation and delivery of electric power. California now contends with wildfires every year, driven by high temperatures and persistent drought conditions. Death tolls from these fires are also on the rise, punctuated by the horrific Camp Fire in 2018, which destroyed 90% of the structures in Butte County, and took the lives of 100 people there.

In all, tens of thousands of structures were destroyed in 2017 and 2018, as thousands of wildfires spanned over 3 million.[1]

While the fires themselves damaged transmission and distribution lines, some of these fires were started by the grid itself. The California Department of Forestry and Fire Protection found that multiple fires were caused by sparks from utility equipment igniting dry tinder brush on the ground.[2] Wildfires will continue to threaten California, but the drought that increases their likelihood has other negative impacts as well.

All fossil fuel, nuclear power, and hydroelectric generation require access to considerable amounts of water to operate. Drought conditions in the American West have become bad enough that hydroelectric dams have reduced power output and have even had to cease operation. In Chapter 1, we discussed the 819-MW generators of Lake Oroville shutting down due to drought. As this book goes to press, similar challenges are faced by Lake Powell, a 1320-MW hydroelectric dam of the Colorado River at Glen Canyon. As water levels reach critical levels, hydrology modeling suggests there is roughly a 1 in 4 chance that it will not be able to produce power by 2024.[3] As hydroelectric sources are threatened by drought, even more pressure will be put on the coal, natural gas, and nuclear plants. These other sources also require huge amounts of water to operate. According to one estimate, electric power generation consumes more than three trillion gallons of water globally per year.[4] All this means that increasing electricity demands will be met with decreasing water supplies, generating chaotic outcomes.

Governments, utilities, and end-users in California are all taking action to address these challenges and improve the state's energy resilience. The California government is responding to challenges of climate change and drought with aggressive programs to shift away from fossil fuel generation. The California Public Utility Commission (CPUC) used an engagement process to understand stakeholder needs and identify barriers to action on resilience. As a result, they adopted rates, tariffs, and rules in 2021 to guide the three largest investor-owned utilities in the state on microgrid deployment in California. This will ensure that regulatory clarity will not stand in the way of microgrid development and the associated resilience benefits. The CPUC also ordered the utilities to create a Microgrid Incentive Program, which would fund clean energy microgrids from a $200 million budget for vulnerable communities impacted by grid outages. The clean energy focus of this program allows the state to leverage one tranche of funding to address two state government goals – energy resilience and climate change mitigation.

California's electric utilities are leveraging a suite of approaches and technologies to, first and foremost, reduce the likelihood of causing a fire and, secondly, to improve their network's resilience. To reduce fire risk, utilities have increased vegetation management, evaluated line reclosers that produce fewer sparks, and are actively deenergizing power lines during times of high wildfire risk. This last measure reduces some major risks but also creates burdens on some communities that are disconnected in what are typically hot months. San Diego Gas & Electric is proposing a series of substation-level microgrids in communities at risk of being deenergized, leveraging local renewables to help power the systems. California utilities are also making major investments in energy storage. By June 2021, California was already leading the nation with 1438 MW of utility-scale battery storage capacity, as much as the next 10 states combined.[5]

California businesses have led the way on energy efficiency, dramatically improving their ability to mitigate peak demand since the turn of the century. End-users have also employed a myriad of on-site generation, energy storage, and microgrid solutions to operate independently from the grid when needed. Many of these end-users participated in virtual power plant programs run by the utilities and earned money by providing energy to the grid during times of need.

In each of these cases, the organizations in question identified energy resilience challenges, engaged stakeholders to understand and evaluate the suite of options, and pursued technical and programmatic solutions best suited to their capabilities and place within the market. Electric utilities could make no moves without support and approval of their government and regulator – which themselves rely on the political will of the people. End-users would be hard pressed to justify efficiency and demand response efforts without utility programs that create a market for those services. Microgrids would have failed to emerge without regulatory clarity around the rules, utility participation in interconnection agreements, and third-party developers to propagate these systems in the marketplace. The efforts of each of these players created or amplified opportunities among the others.

It is clear that all aspects of our society and economy are going to be impacted by threats to the energy system. What is also clear is that, by engaging all parties in a coordinated effort to improve our resilience, our households, companies, cities, utilities, states, and nations have concrete, actionable ways to weather the storm. Hopefully, this book has helped you to understand the challenges and options related to threats to the energy system. Best of luck in developing your plan to move your organization to exactly where it needs to be on the spectrum of energy resilience.

Index

A

Adrienne Lotto, 89, 106, 191
American Reinvestment and Recovery Act, 166
American Society of Civil Engineers, 16, 187
AMI meters, 95
Amory Lovins, 33
Army Corps of Engineers, 123

B

Bennet Chabot, 88, 104
Better Buildings Program, 118
Biden Administration, 22
BJ Tomlinson, 67
Brian Coughlan, 104, 192
Build Back Better Recovery Plan, 33

C

California, 11, 15, 24, 25, 35, 37, 42, 48, 79, 98, 104, 127, 167, 170, 175, 176, 177, 185, 186, 188, 189, 196
California Department of Forestry and Fire Protection, 175
California Integrated Climate Adaptation and Resiliency Program, 37
California Public Utility Commission (CPUC), 25, 176
CenterPoint Energy, 125, 195
Climate Change, 23, 52, 185, 186, 186, 187, 188, 189, 191, 193, 195
Colorado, 15, 176
Community Energy Strategic Planning Academy, 65, 183
Con Edison, 102, 125
Cornell University, 120, 183, 194
COVID-19, 10, 77, 117, 118
Crystall Merlino, 74, 159, 190, 196

D

Damir Novosel, 77, 190
Department of Defense, 49, 114, 115, 116, 193, 193, 193, 195
Department of Energy, 23, 32, 52, 95, 118, 158, 160, 161, 164, 166, 168, 183, 186-191, 193, 195, 196
Department of Homeland Security (DHS), 21, 53, 65, 74, 158, 159, 189, 190, 196
Deregulation, 32
DHS Resilience Framework, 51, 52, 189, 65, 190, 192
Distributed Energy Resources, 86, 192, 102, 121, 194
Dr. Dafni Bar-Sagi, 119
Drought, 25, 175, 186, 188

E

Electrical Advisory Committee, 95
Electric Reliability Council of Texas, 20, 37, 98
El Paso, 67
Energy-as-a-Service, 34, 61, 135, 150
Energy Efficiency, 129, 135, 154, 166, 189, 193
Energy Resilience and Conservation Investment Program (ERCIP), 117
Energy Service Performance Contracts, 59
ENERGY STAR, 132, 166

Energy Storage, 103, 146, 195
Entergy, 1, 19
Environmental Protection Agency (EPA), 168, 183
Erik Svanholm, 73, 88, 90, 169, 190, 191, 196

F

Federal Energy Management Program (FEMP), 69, 189
Federal Energy Regulatory Commission (FERC), 32, 101
Flood, 64, 96, 194, 196
Forest Fire (see 'Wildfire), 196
Fort Bliss, 67, 68
Fort Carson, 116
Fossil Fuels, 9
Fuel Cells, 140, 195

G

Galveston Bay, 125
Gavin Newsome, 11, 185
Government Accountability Office, 10, 99, 185-188, 191
Governor Schwarzenegger, 37
Guide for Climate Change Resilience Planning, 52, 64, 189
Gunderson Health System, 118, 193

H

Hazard Mitigation Grant Program, 121
Hilcorp Liberty project, 11
Hoover Dam, 16
Houston, 3, 74, 119, 123-126, 185, 194, 195
Hunts Point, 125
Hurricane, 1, 2, 19, 27, 28, 36, 45, 117, 119, 120, 121, 123-126, 185, 187, 188, 193, 194
Hurricane Ida, 1, 19, 117, 119, 124, 185, 187, 193, 194
Hurricane Katrina, 1, 2, 19, 124
Hurricane Sandy, 27, 28, 36, 45, 117, 120, 121, 123, 125, 188
Hydraulic Fracking, 10

I

Interconnection, 20, 37, 98

J

Jennifer DeCesaro, 160, 196
Jimmy Carter, 32

K

Keystone Pipeline, 12

L

Lake Mead, 16
Langone Medical Center, 117, 119, 123

M

Marissa Aho, 3, 125, 126
Microgrid, 37, 56, 151, 176, 192, 193, 194, 195
Montgomery County, 122, 123, 194
Multi-year rate plans, 170

N

National Association of Regulatory Utility Commissioners (NARUC), 106, 109, 192
National Association of State Energy Officials (NASEO), 110, 192, 195
National Renewable Energy Laboratory (NREL), 18, 38, 69, 71, 91, 98, 134, 163, 164, 183, 165, 187, 189, 190, 191, 193, 195, 196
National Research Council, 21, 187
National Weather Service, 124, 194
Net-Zero Military Installations planning guide, 49
Nevada, 15, 16
New Jersey, 37, 118, 121-123, 189, 194

New Mexico, 48, 67, 127
New Orleans, 1, 19, 20, 119, 187
New York, xiv, 1, 11, 19, 26, 31, 33, 37, 41, 47, 89, 96, 105, 106, 117, 120, 123-125, 127, 160, 162, 163, 185-189, 191, 193, 194, 196
New York Economic Development Corporation, 125
New York Power Authority, 89, 106, 189, 191
New York State Energy Research and Development Authority (NYSERDA), 37, 163
New York University (NYU), 194
Non-wires alternatives, 20
North American Electric Reliability Council (NERC), 31
NYPrize, 163

O

Obama Administration, 12
Organization of Petroleum Exporting Countries (OPEC), 31

P

Pacific Gas & Electric (PGE) 88, 95, 191, 192
Pacific Northwest National Laboratory (PNNL), 69, 134
Performance-based regulation, 105, 173
Performance Excellence in Electricity Renewal (PEER), 122
Photovoltaics, 143, 144
Photovoltaics (see 'Solar Power'), 181
Power Purchase Agreement (PPA), 59
Public Utility Regulatory Policies Act of 1978, 168

R

Reforming the Energy Vision, 37, 162, 189
Resilience Innovation for a Stronger Economy (RISE), 125
Revolving loan fund, 60
Risk Assessment, 52
Rocky Mountain Institute, 33
Ron Critelli, 91, 191

S

Sabotage, 20, 182
San Diego Gas & Electric+1:147, 176
Saved power purchase agreement, 59
Sharla Artz, 93, 164, 196
Shell Oil, 12
Smart Grid, 88, 190, 191, 101, 192, 192, 193, 194
Smart Grid Investment Grant (SGIG), 166
Smart Power Infrastructure Demonstration for Energy Reliability and Security (SPIDERS), 161 196
Smart Switchgear, 95, 100
Solar Power, 63, 120, 144
Spectrum of Energy Resilience, 5
Stop & Shop, 127, 195
System Average Interruption Duration Index (SAIDI), 189
System Average Interruption Frequency Index (SAIFI), 189

T

Terrorism, 20, 187, 189
Terrorism (see 'Sabotage'), 182
Tesla, 138
Texas, 1, 3, 14, 15, 20, 34, 35, 37, 67, 91, 98, 124, 125, 126, 137, 185, 186, 194, 195
Thermoelectric Power, 13
Tokyo Electric Power Company Holdings Inc. (TEPCO), 97
Toronto Hydro, 95
Transportation, 18, 55, 122
Trump Administration, 11, 12

U

Union of Concerned Scientists, 14, 25, 186
University Medical Center of Princeton, 118
Utility Restructuring, 5, 17, 168

V

Vermont, 36
Vince Guthrie, 116

W

Water, 13, 14, 15, 25, 185, 186, 188, 190, 196
Weather / Storms, 2, 14, 20, 23, 25-27, 77, 108, 126, 140, 124, 185, 187, 194
Wind Power, 144

X

Xcel Energy, 93, 164, 196

About the Authors

Alex Rakow is the Sustainability Lead for the Cloud & Service Provider segment at Schneider Electric, where he works with hyperscale data center operators to advance their sustainability and energy performance. He has 11 years of experience working on sustainability and energy resilience issues with both private and public clients, including the National Park Service and the Environmental Protection Agency. Before coming to Schneider, he was part of the team that created the microgrids business at Hitachi America. He has a BS from Cornell University in Environmental Science, and an MA from Johns Hopkins University focused on global economic development and sustainability. Mr. Rakow is a Certified Energy Manager, and serves on the Climate and Sustainable Energy Advisory Board for his local government in Tompkins County, New York.

Brian Levite is a Certified Energy Manager with 23 years of experience in the energy industry. He has supported the Environmental Protection Agency, Department of Energy, and numerous corporations and electric utilities in areas of energy strategy. While at the National Renewable Energy Laboratory, Mr. Levite was the one of the developers of and instructors for DOE's Community Energy Strategic Planning Academy. He has a BA in Environmental Policy and an MA in Public Policy from The American University. Mr. Levite currently works at ICF Consulting where he helps utility and government clients understand develop resilience plans and incorporate new technologies into the grid.

Endnotes

Introduction

[1] Louisiana Department of Health. Press Release: "LDH: Hurricane Ida Storm-Related Death Toll Rises to 26." September 8, 2021.

[2] Johnson, Krista. "Two Alabama Electric Workers Die Trying to Restore Power Lost During Ida." Montgomery Advertiser. September 1, 2021.

[3] National Centers for Environmental Information. National Oceanic and Atmospheric Administration. "Billion-Dollar Weather and Climate Disasters." Website: https://www.ncdc.noaa.gov/billions/events.

[4] Maria Aho, Policy Director & Chief Resilience Officer for Houston Texas. Interview with Author, conducted on XX XX, 2020.

Chapter 1

[1] US Energy Information Administration, "US Energy Outlook, 2021." February 3, 2021.

[2] Government Accountability Office, "Information on Shale Resources, Development, and Environmental and Public Health Risks." September 2012.

[3] U.S. EPA, "Impacts from the Hydraulic Fracturing Water Cycle on Drinking Water Resources in the United States (Final Report)." 2016.

[4] Concerned Health Professionals of New York, & Physicians for Social Responsibility, "Compendium of Scientific, Medical, and Media Findings Demonstrating Risks and Harms of Fracking (Unconventional Gas and Oil Extraction)." 7th ed. December 2020.

[5] Ogneva-Himmelberger, Yelena and Liyao Huang, "Spatial Distribution of Unconventional Gas Wells and Human Populations in the Marcellus Shale in the United States: Vulnerability Analysis." *Applied Geography*, Vol. 60, 165-174, June 2015.

[6] Office of Governor Gavin Newsome, "Governor Newsom Takes Action to Phase Out Oil Extraction in California." April 23, 2021.

[7] Koenig, Ravenna and Harball, Elizabeth, "Climate Change Slows Oil Company Plan to Drill in the Arctic." NPR. November 21, 2019.

[8] Krauss, Clifford and Stanley Reed, "Shell Exits Arctic as Slump in Oil Prices Forces Industry to Retrench." The New York Times. September 28, 2015.

[9] Fountain, Henry. "Sale of Drilling Leases in Arctic Refuge Fails to Yield a Windfall." The New York Times. January 6, 2021. Updated May 27, 2021.

[10] Government Accountability Office, "CRITICAL INFRASTRUCTURE PROTECTION – Actions Needed to Address Weaknesses in TSA's Pipeline Security Program Management." 2018.

[11] Howarth, Robert W., "Ideas and Perspectives: Is Shale Gas a Major Driver of Recent Increase in Global Atmospheric Methane?," *Biogeosciences*, Vol. 16, 3033–3046, 2019.

[12] Rogers, John, et al., "Water-Smart Power: Strengthening the U.S. Electricity System in a Warming World." Cambridge, MA: Union of Concerned Scientists, July, 2013.

[13] "Energy Demands on Water Resources: Report to Congress on the Interdependency of Energy and Water." US Department of Energy. December, 2006.

[14] Brown, Thomas, et al., "Projected Freshwater Withdrawals in the United States under a Changing Climate," *Water Resources Research*, Vol. 49, 1259-1276, 2013.

[15] Dieter, C.A. et al., Estimated Use of Water in the United States in 2015: U.S. Geological Survey Circular 1441, 65 p. 2018.

[16] Rogers, John, et al., "Water-Smart Power: Strengthening the U.S. Electricity System in a Warming World," Cambridge, MA: Union of Concerned Scientists, July 2013.

[17] Penney, Veronica, "How Texas' Power Generation Failed During the Storm, in Charts." The New York Times. February 19, 2021.

[18] Adams-Heard, Rachel, Javier Blas, and Mark Chediak, "A Giant Flaw in Texas Blackouts: It Cut Power to Gas Supplies." Bloomberg.com. February 20, 2021.

[19] Penney, Veronica, "How Texas' Power Generation Failed During the Storm, in Charts." The New York Times. February 19, 2021.

[20] Oxner, Reese, "Texans Now Face a Water Crisis After Enduring Days Without Power." The Texas Tribune. February 19, 2021.

[21] "Effects of Climate Change on Energy Production and Use in the United States." U.S. Climate Change Science Program and the Subcommittee on Global Change Research. February 2008.

[22] DiSavino, Scott, "California Drought Cuts Hydropower, Boosts Natgas Prices." Reuters. May 21, 2021.

[23] Meeks, Alexandra, "A California Reservoir is Expected to Fall So Low that a Hydro-Power Plant will Shut Down for First Time." CNN. June 17, 2021.

[24] Kenny, J. F., et al., "Estimated Use of Water in the United States in 2005." U.S. Geological Survey Circular. 1344, October 2009.

[25] Rogers, John, et al., "Water-Smart Power: Strengthening the U.S. Electricity System in a Warming World," Cambridge, MA: Union of Concerned Scientists, July 2013.

[26] "2021 Infrastructure Report Card - Energy." American Society of Civil Engineers. 2021.

[27] "Failure to Act – The Impact of Current Infrastructure Investment on America's Economic Future." American Society of Civil Engineers. 2012.

[28] Amin, Massoud, "Continuing Crises in National Transmission Infrastructure: Impacts and Options for Modernization." National Science Foundation. June 2004.

[29] Amin, Massoud, "Turning the Tide on Outages – What are the True Costs of Implementing – or Failing to Implement – a Stronger, Smarter and More Robust Grid." DRAFT. July 18, 2011.

[30] US EPA, "Inventory of U.S. Greenhouse Gas Emissions and Sinks." 1990-2019. 2021.

[31] Mai, Trieu, et al., "Electrification Futures Study: Scenarios of Electric Technology Adoption and Power Consumption for the United States." Golden, CO: National Renewable Energy Laboratory. 2018.

[32] Kasakove, Sophie and Nicholas Bogel-Burroughs, "New Orleans Built a Power Plant to Prepare for Storms. It Sat Dark for 2 Days." The New York Times. September 10, 2021.

[33] Eavis, Peter and Ivan Penn, "Why Louisiana's Electric Grid Failed in Hurricane Ida." The New York Times. September 17, 2021.

[34] "The Value of Increased HVDC Capacity Between Eastern and Western U.S. Grids: The Interconnections Seam Study." National Renewable Energy Laboratory. October 2020.

[35] "Terrorism and the Electrical Power Distribution System." National Research Council, National Academy of Sciences. 2012.

[36] "Terrorism and the Electrical Power Distribution System." National Research Council, National Academy of Sciences. 2012.

[37] Nicole Perlroth, "Russian Hackers Targeting Oil and Gas Companies." The New York Times. June 30, 2014.

[38] Sanger, David E. and Nicole Perlroth, "Pipeline Attack Yields Urgent Lessons About U.S. Cybersecurity." The New York Times. May 14, 2021. Updated June 8, 2021.

[39] "U.S. Energy Sector Vulnerabilities to Climate Change and Extreme Weather." U.S. Department of Energy. July 2013.

[40] Lindsey, Rebecca, and LuAnn Dahlman, "Climate Change: Global Temperature." National Oceanographic and Atmospheric Administration. March 15, 2021.

[41] "Report to Congressional Requesters – Climate Change – Energy Infrastructure Risks and Adaptation Efforts." U.S. Government Accountability Office. January 2014.

42 Reidmiller, D. R., C. W. Avery, D. R. Easterling, K. E. Kunkel, K. L. M. Lewis, T. K. Maycock, and B. C. Stewart (eds.). *Impacts, Risks, and Adaptation in the United States: Fourth National Climate Assessment, Volume II*. U.S. Global Change Research Program. 2018.

43 Remer, White, Sustainability & Resilience Officer, City of Tampa. Interview with the author. Conducted. XX XX

44 Fecht, Sarah, "How Climate Change Impacts Our Water." State of the Planet. Columbia University. September 23, 2019.

45 U.S. Drought Monitor, "California – Thurs. August 4, 2021." 2021. droughtmonitor.unl.edu/. Accessed August 5, 2021.

46 "Report to Congressional Requesters – Climate Change – Energy Infrastructure Risks and Adaptation Efforts." U.S. Government Accountability Office. January 2014. http://www.gao.gov/assets/670/660558.pdf.

47 "How Hurricane Sandy Affected the Fuels Industry." NACS Online. 2013. Accessed May 15, 2014.

48 "Report to Congressional Requesters – Climate Change – Energy Infrastructure Risks and Adaptation Efforts." U.S. Government Accountability Office. January 2014.

49 Bradbury, James, et. al., "Climate Change and Energy Infrastructure Exposure to Storm and Sea-Level Rise." Oak Ridge National Laboratory, Office of Energy Policy and Systems Analysis, US Department of Energy. July 2015.

Chapter 2

1 "The New York Times: Our history / 1965." *The New York Times*. 2020.

2 North American Energy Reliability Council. "Milestones: NERC Reliability Standards." May 19, 2014.

3 US Energy Information Administration. "Annual energy review – total energy." November, 2020.

4 Kenneth Dubin. "Oil-fired power plants provide small amounts of U.S. electricity capacity and generation." US Energy Information Administration. May, 2017.

5 Minkel, JR. "The 2003 Northeast Blackout – Five Years Later." *Scientific American*. August 13, 2008.

6 Biden, Joseph R. "Remarks by President Biden Before Signing Executive Actions on Tackling Climate Change, Creating Jobs, and Restoring Scientific Integrity." The White House. January 27, 2021.

7 Morehouse, C. "FERC closes resilience docket opened in response to DOE coal, nuclear bailout proposal." *Utility Dive*. February 19, 2021.

8 Englund, W. "The grid's big looming problem: Getting power to where it's needed." *The Washington Post*. June 29, 2021.

9 New York Power Authority. "Reforming the Energy Vision for New York." https://www.nypa.gov/innovation/initiatives/rev. 2021.

10 New York-New Jersey CHP Technical Assistance Partnerships. "Program profile – New Jersey energy resilience bank." 2019.

11 State of California – Governor's Office of Planning and Research. "Impact report and 2020 program recommendations." April, 2020.

12 "The value of increased HVDC capacity between Eastern and Western U.S. grids: The interconnections seam study," National Renewable Energy Laboratory, published in IEEE Transactions on Power Systems, September 2021.

13 Andrew Z, "Op-ed: Learning to bounce back." *The New York Times*. November 2, 2012.

Chapter 3

1 US Department of Homeland Security. "DHS Resilience Framework: Providing a Roadmap for the Department in Operational Resilience and Readiness." 2018.

2 US Department of Energy, Office of Energy Policy and Systems Analysis. *"Climate Change and the Electricity Sector: Guide for Climate Change Resilience Planning."* September 2016.

3 Castro, Chris. Director of Sustainability & Resilience. City of Orlando, Florida. Interview with the author. Conducted 3/8/21.

4 US Department of Energy, Office of Energy Efficiency & Renewable Energy. "Guide to Community Energy Strategic Planning." March 2013.

5 US Department of Energy, Federal Energy Management Program. *"Technical Resilience Navigator."* 2020.

6 Remer, Whit. Sustainability & Resilience Officer for the City of Tampa. Interview with author. Conducted 3/1/21.

Chapter 4

1 "Valuing Resilience in Electricity Systems (nrel.gov) Resilience in Electricity Systems," National Renewable Energy Laboratory, September 2019, https://www.nrel.gov/docs/fy19osti/74673.pdf

2 System average interruption duration index (SAIDI) and system average interruption frequency index (SAIFI) are commonly used by electric utilities and regulators to understand energy reliability performance.

3 "Resilience Framework, Methods, and Metrics for the Electricity Sector," IEEE Power & Energy Society, October 2020, https://resourcecenter.ieee-pes.org/publications/technical-reports/PES_TP_TR83_ITSLC_102920.html

[4] Interview with Erik Svanholm, Vice President of Non-Wires Alternatives, S&C Electric Co., 2/17/21.

[5] DHS Resilience Framework | Homeland Security, https://www.dhs.gov/publication/dhs-resilience-framework

[6] Interview with Crystall Merlino, Director, Resilience and Energy Management, U.S. Department of Homeland Security, 3/4/21.

[7] "Valuing Resilience in Electricity Systems," National Renewable Energy Laboratory, September 2019, https://www.nrel.gov/docs/fy19osti/74673.pdf

[8] "Designing Policy Roadmap," Ted Ko, Tyler Wakefield, STEM, https://www.stem.com/designing-policy-roadmap-to-clean-resilient-grid/

[9] Interview with Damir Novosel, President at Quanta Technology, 4/21/21.

[10] "Grid Modernization: Metrics Analysis (GMLC1.1) – Resilience," Grid Modernization Laboratory Consortium, April 2020, https://gmlc.doe.gov/sites/default/files/resources/GMLC1.1_Vol3_Resilience.pdf

[11] "Grid Modernization: Metrics Analysis (GMLC1.1) – Resilience," Grid Modernization Laboratory Consortium, April 2020.

[12] "Resilience Framework, Methods, and Metrics for the Electricity Sector," IEEE Power & Energy Society, October 2020, https://resourcecenter.ieee-pes.org/publications/technical-reports/PES_TP_TR83_ITSLC_102920.html

Chapter 5

[1] "2020 State of Commercial & Industrial Power Reliability Report," S&C Electric Company, January 2021, https://www.sandc.com/en/solutions/reliability/#2021%20C&I%20Report

[2] "Electric Vehicle Outlook 2020," Bloomberg New Energy Finance.

[3] Sources of Greenhouse Gas Emissions | Greenhouse Gas (GHG) Emissions | U.S. EPA.

[4] 1/20/21 press release from the Edison Electric Institute.

[5] Internal, non-published, S&C research used in this book with permission from the company.

[6] For a more in-depth discussion of the benefits of the smart grid, see "How the Smart Grid Promotes a Greener Future," U.S. Department of Energy, https://www.energy.gov/sites/prod/files/oeprod/DocumentsandMedia/Environmentalgroups.pdf, and "Estimating the Costs and Benefits of the Smart Grid," Electric Power Research Institute, March 2011, https://www.smartgrid.gov/files/documents/Estimating_Costs_Benefits_Smart_Grid_Preliminary_Estimate_In_201103.pdf

[7] Interview with Erik Svanholm, Vice President of Non-Wires Alternatives, S&C Electric Co., 2/17/21.

[8] Interview with Bennett Chabot, Product Manager, Grid Innovation, Pacific Gas & Electric, 4/19/21.

[9] Interview with Adrienne Lotto, VP & Chief Risk and Resilience Officer at New York Power Authority, 4/22/21.

[10] Interview with Erik Svanholm, Vice President of Non-Wires Alternatives, S&C Electric Co., 2/17/21.

[11] Interview with Ron Critelli, Senior Director - Power Delivery Engineering & Technical Services at Florida Power & Light, 3/19/21.

[12] "Power Sector Resilience Planning Guidebook: A Self-Guided Reference for Practitioners," Stout, Lee, Cox, Elsworth, and Leisch, https://www.nrel.gov/docs/fy19osti/73489.pdf, June 2019.

[13] Interview with Ron Critelli, Senior Director - Power Delivery Engineering & Technical Services at Florida Power & Light, 3/19/21.

[14] "Utility Vegetation Management Best Practices," Article by Davey Resources Group on The Utility Expo Web Site, https://www.theutilityexpo.com/news/utility-vegetation-management-best-practices

[15] "Big Data Analytics: Recommendations for the U.S. Department of Energy," DOE Electricity Advisory Committee, February 2021, https://www.energy.gov/sites/prod/files/2021/02/f82/EAC%20Big%20Data%20Analytics%20Work%20Product_Final.pdf

[16] Interview with Bennett Chabot, Product Manager, Grid Innovation, Pacific Gas & Electric, 4/19/21.

[17] "Rethinking Smart Grid Data Analytics," T&D World, August 24, 2014, https://www.tdworld.com/smart-utility/article/20964697/rethinking-smart-grid-data-analytics

[18] "Flooding and Climate Change: Everything You Need to Know," Melissa Denchak, Natural Resources Defense Council Web Site, April 10, 2019, https://www.nrdc.org/stories/flooding-and-climate-change-everything-you-need-know

[19] Interview with Ron Critelli, Senior Director - Power Delivery Engineering & Technical Services at Florida Power & Light, 3/19/21.

[20] "The Value of Increased HVDC Capacity Between Eastern and Western U.S. Grids: The Interconnections Seam Study," National Renewable Energy Laboratory journal article pre-print, October 2020, https://www.nrel.gov/docs/fy21osti/76850.pdf

[21] O'Neill, Patrick Howell. "Russian hackers tried to bring down Ukraine's power grid to help the invasion." MIT Technology Review. April 12, 2022.

[22] "Critical Infrastructure Protection: Actions Needed to Address Significant Cybersecurity Risks Facing the Electric Grid," United States Government Accountability Office, August 2019, https://www.gao.gov/assets/gao-19-332.pdf

[23] "State of the Electric Utility 2021: Utilities' cybersecurity approach shows cause for concern, experts say," Utility Dive Web Site, April 1, 2021, https://www.utilitydive.com/news/state-of-the-electric-utility-2021-utilities-cybersecurity-approach-shows/596664/

[24] "Review of Demand Response under Smart Grid Paradigm," Balijepalli *et al.*, IEEE PES Innovative Smart Grid Technologies, December 2011, https://ieeexplore.ieee.org/abstract/document/6145388

[25] "BQDM Program Demonstrates Benefits of Non-Traditional Utility Investments," Coley Girouard, UtilityDive, 3/11/19, https://www.utilitydive.com/news/bqdm-program-demonstrates-benefits-of-non-traditional-utility-investments/550110/

[26] "S&C Builds Ameren a Microgrid To Study Distribution Use Cases," S&C Electric Co. website 6/25/18, https://www.sandc.com/globalassets/sac-electric/documents/sharepoint/documents---all-documents/case-study-180-1076.pdf?dt=637665423181378717

[27] Interview with Bennett Chabot, Product Manager, Grid Innovation, Pacific Gas & Electric, 4/19/21.

[28] For a more fulsome review of microgrid market regulatory barriers, see "State Regulatory and Policy Considerations for Increased Microgrid Deployment," Prepared by Energetics Inc. for the National Electrical Manufacturers Association, January 2018, https://www.nema.org/docs/default-source/blogdocuments/standards/complimentarydocuments/mgrd-20r2-2018-20contents-20and-20scope.pdf

[29] Interview with Brian Coughlan, President of Utility Management Services, Inc., 2/12/21.

[30] "The Value of Resilience for Distributed Energy Resources: An Overview of Current Analytical Practices," Prepared by Converge Strategies for the National Association of Regulatory Utility Commissioners, April 2019, https://pubs.naruc.org/pub/531AD059-9CC0-BAF6-127B-99BCB5F02198

Chapter 6

[1] Rickerson, Wilson, Kiera Zitelman, and Kelsey Jones, "Valuing Resilience for Microgrids: Challenges, Innovative Approaches, and State Needs." National Association of Regulatory Utility Commissioners. National Association of State Energy Officials. February, 2022.

[2] Beck, Fred and Eric Martinot, "Renewable Energy Policies and Barriers," in *Encyclopedia of Energy*, edited by Cleveland Culter (San Diego: Academic Press/Elsevier Science, 2004), 365-383.

3 Beck, Fred and Eric Martinot, "Renewable Energy Policies and Barriers," in *Encyclopedia of Energy*, edited by Cleveland Culter (San Diego: Academic Press/ Elsevier Science, 2004), 365-383.

4 Rappaport, Ann and Sarah Creighton, *Degrees That Matter: Climate Change and the University* (Boston: MIT Press, 2007).

5 U.S. Department of Defense, "Department of Defense Annual Energy Management and Resilience Report, Fiscal Year 2020." Office of the Assistant Secretary of Defense for Sustainment. September, 2020.

6 Van Broekhoven, S. B., et al., "Microgrid Study: Energy Security for DoD Installations, Technical Report of Lincoln Laboratory," Massachusetts Institute of Technology, June 18, 2012.

7 Risser, Roland, "Energy Department's Hospital Energy Alliance Helps Partner Save Energy and Money." U.S. Department of Energy. September 4, 2012.

8 "Metrics and Standards for Energy Resilience at Military Installations." Memorandum. The Office of the Under Secretary of Defense. United States Department of Defense. May 20, 2021.

9 Surash, J.E. Jack and Mike McGhee, U.S. Army. "Resourcing Installations Energy Resilience Solutions." Press Release. April 13, 2020.

10 Jaffe, Mark, "Army Enlisting Help to Make its Bases Energy Independent." The Denver Post. June 3, 2012.

11 Risser, Roland, "Energy Department's Hospital Energy Alliance Helps Partner Save Energy and Money." U.S. Department of Energy. September 4, 2012.

12 Heiman, Michael, "What Caused Generators to Fail at NYC Hospitals?" CBS News. October 28, 2012.

13 Hartocollis, Anemona, "A Flooded Mess That Was a Medical Gem." The New York Times. November 9, 2012.

14 Achenbach, Joel, Akilah Johnson, and Jacqueline Dupree, "Hurricane Ida Forces Three Damaged Hospitals to Evacuate Patients." The Washington Post. August 30, 2021.

15 Gunderson Health System, "Envision." https://www.gundersenenvision.org/envision.

16 Beiermeister, Mike, "Gundersen's Sparta Clinic Receives National Praise for Energy Efficiency." WKXO. July 23, 2019.

17 Rich, Jeff, Executive Director, Gunderson Envision Program, Gundersen Health System, Interview with author, conducted May 21, 2013.

18 Rabner, Barry, President and CEO, Princeton HealthCare System, Interview with author, conducted May 21, 2013.

19 Sifferlin, Alexandra, "Storm Destroys Valuable Medical Research." Time Magazine. November 2, 2012.

[20] Britto, Brittany, "Hurricane Ida Forces Tulane University to Evacuate Students to Houston." The Houston Chronicle. August 31, 2021.

[21] Joyce, Lanny, Director of Energy Management, Cornell University, Interview with author, conducted July 30, 2013.

[22] Cornell University – Sustainable Campus. "Solar Renewable Energy." https://sustainablecampus.cornell.edu/campus-initiatives/buildings-energy/solar-energy/

[23] Pyper, Julia, "Are Microgrids the Answer to City-Disrupting Disasters?" Scientific American. September 11, 2013.

[24] New York University, "NYU Switches on Green CoGen Plant and Powers Up for Sustainable Future." Press Release. January 21, 2011.

[25] Office of the Governor, State of New Jersey, "Christie Administration Announces New Hazard Mitigation Initiative to Support Post-Sandy Energy Resilience." Press Release. July 10, 2014.

[26] New Jersey Board of Public Utilities, "Microgrid Report." 2016.

[27] Wood, Elisa, "Not Wanting to Become Texas, New Jersey Grants $4 Million for Microgrid Studies." Microgrid Knowledge. March 12, 2021.

[28] Wood, Elisa, "Montgomery County Microgrids Now Live and Leading." Microgrid Knowledge. October 26, 2018.

[29] Bennett, Sharon, "Montgomery County's Microgrid Electric Bus Depot Under Construction." Microgrid Knowledge. 2021.

[30] Wolfe Emma, et al., "The New Normal: Combating Storm-Related Extreme Weather in New York City." Office of the Deputy Mayor for Administration. September, 2021.

[31] Barnard, Anne, et al., "At Least 43 are dead After Ida Causes Flooding in Four States." The New York Times. September 2, 2021.

[32] Lee, April and Tyler Hodge, "Hurricane Harvey Caused Electric System Outages and Affected Wind Generation in Texas." Energy Information Administration. September 13, 2017.

[33] Carroll, Susan, "Nature Ruled, Man Reacted. Hurricane Harvey was Houston's Reckoning." The Houston Chronicle. December 7, 2021.

[34] Blake, Eric S. and David A. Zelinsky, "National Hurricane Center Tropical Cyclone Report – Hurricane Harvey." National Weather Service. May 9, 2018.

[35] City of Houston. Website: "Post Harvey – Flood Mitigation Projects." http://www.houstontx.gov/postharvey/flood-mitigation-projects.html

[36] Aho, Marissa, Policy Director and Chief Resilience Officer, City of Houston, Texas, Interview with author, conducted February 3, 2021.

[37] Maloney, Peter, "Small Businesses in New York City Install Microgrids with RISE Funds." Microgrid Knowledge. July 23, 2019.

[38] Hosfelt, Bailey, "What Climate Risks Mean for NYC's Food Supply." CityLimits. December 30, 2019.

[39] Maloney, Peter, "Latest Electrificaiton Project at JFK Airport Completed." Microgrid Knowledge. October 2, 2019.

[40] Shapiro, Emily, Max Golembo, Melissa Griffin, and Ivan Pereira, "Nicholas Slams Gulf Coast with Dangerous FLOODING: LATEST FORECAST." ABC News. September 16, 2021.

[41] City of Houston Texas, Office of the Mayor. "City of Houston and CenterPoint Energy Announce Transformative Initiative to Enhance Energy Resilience and Promote Transition to Sustainable Energy." Press Release. February 9, 2022.

[42] Wood, Elisa, "Stop & Shop, Major Northeast Grocery Store Chain, to Install 40 Microgrids." Microgrid Knowledge. January 14, 2020.

[43] Howard, Ethan, "Blackstone Commits $500 Million for GreenStruxure Microgrids." Microgrid Knowledge. May 25, 2021.

[44] St. John, Jeff, "Goldman Sachs and Shell Bet $75M That Chain Stores Can Boost Grid Resiliency." Canary Media. March 7, 2022.

[45] Masanet, Eric, et al., "Recalibrating Global Data Center Energy-Use Estimates." Science. Vol. 367, No. 6481. pp. 9884-986. February 28, 2020.

[46] Behling, Noriko, "Making Fuel Cells Work," *Issues in Science and Technology*, Issue 29-3, 12, 2013.

[47] Skok, Andy, "Fuel Cell CHP for Industrial & Critical Infrastructure Support," *Presentation at the World Energy Engineering Congress*, October 2, 2014.

[48] Feldman, David, et al., "U.S. Solar Photovoltaic System and Energy Storage Cost Benchmark: Q1 2020." National Renewable Energy Laboratory. January 2021.

[49] Wiser, Ryan, Mark Bolinger, et al., "Land-Based Wind Market Report: 2021 Edition." U.S. Department of Energy – Lawrence Berkeley National Laboratory. August, 2021.

[50] Henze, Veronika, "Battery Pack Prices Cited Below $100/kWh for the First Time in 2020, While Market Average Sits at $137/kWh." Bloomberg New Energy Finance. December 16, 2020.

[51] Ervin, Grant, Chief Resilience Offer, City of Pittsburgh, Interview with author, conducted April 29, 2021.

Chapter 7

[1] October 7, 2021 Announcement from the White House, https://www.whitehouse.gov/briefing-room/statements-releases/2021/10/07/fact-sheet-biden-administration-releases-agency-climate-adaptation-and-resilience-plans-from-across-federal-government/

[2] Interview with Crystall Merlino, Director, Resilience, Sustainability, and Energy Management, Department of Homeland Security, 3/4/21.

[3] Interview with Jennifer DeCesaro, Director, Recovery and Resilience at U.S. Department of Energy, 2/26/21.

[4] Technology Transition Final Public Report: Smart Power Infrastructure Demonstration for Energy Reliability and Security, December 2015, https://www.energy.gov/sites/prod/files/2016/03/f30/spiders_final_report.pdf

[5] Interview with Sharla Artz, Director, Security and Resilience Policy, Xcel Energy, 4/19/21.

[6] https://www.nrel.gov/security-resilience/energy-resilience.html

[7] "Accelerating Clean Energy Commercialization: A Strategic Partnership Approach", National Renewable Energy Laboratory Technical Report, NREL/TP-6A60-65374, April 2016.

[8] https://www.energy.gov/oe/information-center/recovery-act

[9] Interview with Erik Svanholm, Vice President of Non-Wires Alternatives, S&C Electric Co., 2/17/21.

Conclusion

[1] "California Wildfires: Key Recommendations to Prevent Future Disasters," McQuillan, Carol Park, Summers, and Dwyer, Independent Institute, June 25, 2019.

[2] "Utility's Power Lines Caused Huge 2019 California Wildfire," AP News, July 16, 2020, as published on https://apnews.com/article/3d35a822b32a447b665281d0bb2e8fd2

[3] "Lake Powell Hits Historic Low, Raising Hydropower Concerns," Metz and Fonseca, Associated Press, March 16, 2022.

[4] "The Water Consumption of Energy Production: An International Comparison," Spang, Moomaw, Gallageher, Kishen, and Marks, Environmental Research, October 8, 2014.

[5] "Inside Clean Energy: In California, the World's Largest Battery Storage System Gets Even Larger," Dan Gearino, Inside Climate News, September 2, 2021.